T0272018

The Role of the Coincidence Site Lattice
in Grain Boundary Engineering

The Role of
the Coincidence Site Lattice in
Grain Boundary Engineering

Valerie Randle

Department of Materials Engineering
University of Wales, Swansea

CRC Press
Taylor & Francis Group
Boca Raton London New York

CRC Press is an imprint of the
Taylor & Francis Group, an **informa** business

Book 659

CRC Press
Taylor & Francis Group
6000 Broken Sound Parkway NW, Suite 300
Boca Raton, FL 33487-2742

© The Institute of Materials 1996

CRC Press is an imprint of Taylor & Francis Group, an Informa business

No claim to original U.S. Government works

ISBN-13: 978-1-86125-006-3 (hbk)

Publisher's Note

The publisher has gone to great lengths to ensure the quality of this reprint but points out that some imperfections in the original copies may be apparent.

Typeset by Fakenham Photosetting

**Visit the Taylor & Francis Web site at
http://www.taylorandfrancis.com**

**and the CRC Press Web site at
http://www.crcpress.com**

Contents

1. Introduction

The coincidence site lattice (CSL) model was first propounded several decades ago.[1] Since then it has become a cornerstone of grain boundary structural research, particularly in cubic materials. Briefly, if the lattices of two grains were allowed notionally to interpenetrate, certain combinations of orientation relationship between the two lattices would result in a periodic array of coinciding sites – an abstraction which becomes physically real at a grain boundary. A boundary with a high density of coincidence points implies 'good fit' of adjacent grains with a concomitant modification of properties such as diffusivity, energy, mobility. The reciprocal density of CSL points is denoted by Σ (see Chapter 2 for a full description of CSL nomenclature and geometry.)

The widespread acceptance of the CSL model as a tool for grain boundary analysis derives mainly from three sources:

- *The CSL appeals because it is a simple geometrical concept.* The idea of a periodic structure at a grain boundary, invoking a better fit between two adjacent lattices, can be visualised relatively easily, particularly with the aid of two-dimensional diagrams.
- *The CSL model is straightforward to apply in practice.* Computer programs can provide routinely a Σ-value from measurements of the orientation of neighbouring grain pairs. These orientations are in turn obtained using diffraction techniques, e.g. transmission electron microscopy (TEM), electron back scatter diffraction (EBSD) in an scanning electron microscope (SEM).
- *There was notable early success in linking certain grain boundary properties with low-Σ CSLs.* In the late 1950s a set of experiments which have come to be regarded as classic demonstrated that in zone-refined lead doping with <4% tin drastically reduced the mobility of non–CSL (i.e. 'general') boundaries yet CSLs remained mobile because their good-fit structure did not permit segregation of the dopants.[2] Among the other experiments which demonstrated unequivocally that CSLs form a property subset were the series of 'sphere-on-plate' rotation experiments by various workers,[3] which showed that grain boundary energy minima correspond to the CSL geometry.

Further successes in applying the CSL as a classification device or correlated with 'special' boundary properties followed and was extended to polycrystals. (Boundary properties, or boundary geometry, are referred to as *special* if they are markedly different from average). On the other hand, the

CSL model did not prove univerally associative with special boundary properties. This was explained by realising that the CSL is a purely *geometrical* model, i.e. the atoms are viewed as hard spheres, and as such has limitations since grain boundary properties also depend on other factors such as interatomic forces and extrinsic dislocation content.[4,5] For example, energies of the same CSLs in several fcc materials are different, yet from a geometrical standpoint these boundaries are identical.[6]

There followed a certain amount of disenchantment with the CSL model during the period when its limitations first emerged. Several papers were written both in this vein and attempting to re-establish the validity of the CSL.[7,8] However, there remained many convincing demonstrations that grain boundaries which were categorised as low-Σ CSLs corresponded to special properties, and so the CSL approach retained its popularity in grain boundary research.[9,10] The major baffling problem was why some boundaries were seen to violate the low-Σ/good properties criterion. Nowadays we know that part of the answer to this question is that the specification of a CSL relationship between the lattices of two grains does not define the actual position of the grain boundary surface itself. The most commonly observed instance depicting the role of the boundary surface (usually called the boundary 'plane') is that of the coherent and incoherent facets on an annealing twin in a low stacking-fault energy material. In CSL notation, a twin is described as $\Sigma 3$, and the coherent and incoherent facets lie on different planes within this CSL. The properties of the two facets are markedly different despite that they are CSLs having the same Σ-value. Figure 1.1 depicts this by showing different levels of precipitation on coherent/incoherent twin facets.

In the early 1980s two things happened which radically changed the course of grain boundary research. The first of these was that the initiative of *'grain boundary engineering'* or *'grain boundary design'* was proposed.[11] Grain boundary engineering is the exploitation of boundary populations with superior properties in order to improve the overall properties of the material such as strength or corrosion resistance. These special boundaries were assumed to be synonomous with CSLs. Implicit in the grain boundary engineering concept is the notion that the proportion of special boundaries in a material is not fixed, but can be controlled. Since grain boundary engineering was first proposed in 1984 there have been some convincing demonstations of its application, but progress towards understanding the fundamental aspects which underpin it is proceeding at a slower rate.

Recently, the commercial viability of grain boundary engineering has started to be accepted and implemented. The most notable success has been

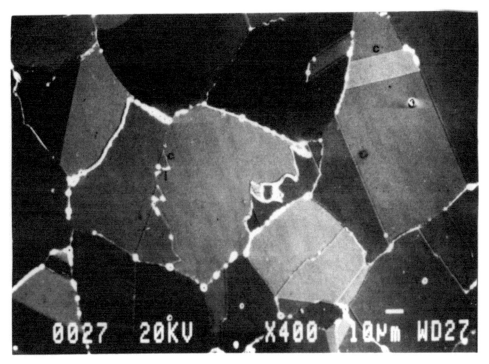

Figure 1.1 Precipitation on grain boundaries in an austenitic steel. The density of precipitation varies, and in particular coherent twins are denuded (e.g. labelled C) whereas incoherent twin facets support precipitates (e.g. labelled I) (Courtesy of M. Caul).

that processing technologies for GBE[TM†] wrought austenitic steels and high performance nickel alloys have been incorporated into production at Ontario Hydro. The industrial process guarantees to produce material with the proportion of special boundaries enhanced to >60%. In turn, this provides increased protection against intergranular corrosion, stress corrosion cracking, embrittlement and sensitisation thus extending service reliability. Figure 1.2 shows the improvements achieved by increasing the proportion of special boundaries in Alloy 600.[12] Most of the special boundaries are in fact twins ($\Sigma3$ in CSL notation) or twin-related boundaries.

The second milestone which has impinged dramatically on grain boundary research is the advent of superior experimental techniques which greatly facilitate the measurement of grain boundary geometry in bulk polycrystals. The most notable advance has been the development of EBSD,[13] which is applied in an SEM and therefore requires minimal specimen preparation. The biggest advantage of EBSD is its efficiency, since the orientations of grains are obtained on-line using a computer algorithm. TEM-based

† GBE[TM] is a registered trademark of Ontario Hydro.

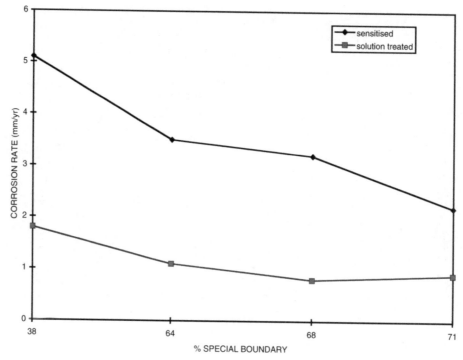

Figure 1.2   Proportions of 'special' boundaries ($\Sigma \leq 29$) in a high performance nickel-based alloy as a function of grain boundary corrosion.[12]

techniques have also improved in recent years, with some machines featuring on-line Kikuchi pattern analysis,[14] and high voltage electron microscopy (HVEM)[15] contributing to grain boundary research. These advances in experimental techniques for measurement of grain boundary parameters, coupled with the grain boundary engineering initiative, have stimulated an increased amount of research in this area.

One of the most recent and exciting development in grain boundary (and microtexture) research methodology is automated capture and indexing of an EBSD pattern, linked to microscope stage and/or beam control. Using fully automated EBSD many thousands of boundaries can be analysed, thus over-coming the traditional problem of small sample populations which may not provide statistically valid representations. Crystal Orientation Mapping (COM) or Orientation Imaging Microscopy (OIM) has developed from automated EBSD, and refers to the mapping of grain boundaries from infor-mation on the change in orientation.[16] CSL designations can be recorded on the image and/or outputted statistically.

At the present time the status of the CSL model is that it largely underpins the resurgence of resource investment in grain boundary research which has been stimulated by the grain boundary engineering initiative. This is not to say that all the discrepancies concerning the CSL model have been resolved, but rather that the convenience and relevance in the majority of cases has turned out to be the overriding factor. In recent years there has been an increase in reported grain boundary investigations using the CSL model, especially in polycrystals, and furthermore the inception of automated EBSD promises to take this research into a new era. Hence it is timely to appraise and consolidate the role that the CSL model has played in grain boundary research, with a view to answering the question *'How valid and useful is the CSL model in the study of grain boundary structure/property relationships in polycrystals?'*.

This, then, is the background against which this major review is set. The objective of it is to provide a critique of the CSL-based investigations which form the sum of our present knowledge, with the main emphasis centring on work from the last two decades or so involving cubic polycrystals. Selection of this timeframe reflects the acceleration in reported investigations over the period. The outcome of this analysis will aim to summarise the progress that has been made using the CSL model and to provide indications of where future work should be directed.

The scheme of the book is as follows. It is first necessary, for the non-specialist, to describe the framework for grain boundary geometry, which forms Section 2.1 in Chapter 2. This is followed by a summary of the formalism of the CSL – insofar as is necessary to understand the results of investigations – in Section 2.2. A description and categorisation of grain boundary properties is dealt with in a similar manner in Section 2.3. Much of the early work concerning CSLs was performed on bicrystals, which underpins our current knowledge. In Chapter 3 this work is linked to the strategies for work on polycrystals, where grain boundary engineering is applicable. An overview of experimental techniques is found in Section 3.3 – again briefly since more detail is available in other texts.

The analysis part of the book begins in Chapter 4. Here statistics from a large body of data taken from most of the readily available sources are presented and discussed, accompanied by consideration of correlations between the CSL and boundary phenomena. In Chapter 5 specific considera-tion is then given to extensions of the CSL, firstly to the important topic of the role of the grain boundary plane in the CSL approach. Both theoretical considerations and experimental data are presented and discussed in Section 5.1. The extension of CSL analyses to connectivity between boundaries is found in Section 5.2. Finally, specific case studies where the CSL model has been applied, and overall conclusions on the role of CSLs in grain boundary engineering comprise Chapter 6, Sections 6.1 and 6.2 respectively.

2. Introduction to Coincidence Site Lattice Boundaries – Geometry and Properties

2.1 BASIC CONCEPTS

2.1.1 Introduction

In this Chapter fundamental aspects of grain boundary geometry, which form the framework for the CSL formalism, will be described for the non-specialist in the field. The treatment is brief, but necessary so that the terminology associated with the application of CSLs which is used throughout this book can be understood on an elementary level without the need for reference elsewhere. The term 'grain boundary geometry' refers to the crystallographic relationship between grains residing in the same polycrystal, usually neighbouring grains.[17] This Chapter forms a summary of the CSL model for grain boundary geometry in cubic polycrystals.

2.1.2 Degrees of freedom

Eight degrees of freedom, five macroscopic and three microscopic, describe a pair of grains joined at a boundary.[18,19] These do not include the relationship between the boundary and external environment of the specimen. The macroscopic degrees of freedom specify two directions and one angle, and characterise completely the overall orientation change between two grains at a grain boundary. This orientation change includes the crystallographic orientation relationship between the lattices of the neighbouring grains, usually called the *misorientation* between grains, and the crystallographic orientation of the grain boundary surface itself. The latter is referred to as the boundary *plane*, despite that under some circumstances it may not be planar.[20,21] For example, the boundaries of newly recrystallised grains are curved, and it is this curvature which subsequently provides the driving force for normal grain growth.

As grain growth proceeds the boundaries become macroscopically more planar. Figure 2.1 illustrates planar boundaries revealed in an aluminium alloy after the grains have been separated by intergranular embrittlement. A

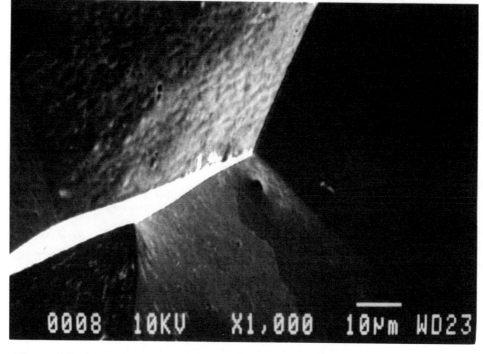

Figure 2.1   Near-planar grain boundaries in an aluminium alloy where the grains have been partially separated by chemical attack.

grain boundary plane such as the planar types shown in Figure 2.1 requires two independent variables to describe its orientation with respect to one grain, as discussed in the next two subsections. Grain boundaries may also feature discontinuities such as facets or, on an atomic scale, ledges as in the coincidence-ledge-dislocation model.[22] The orientation of grain boundary planes forms the topic of Section 5.1.

The three microscopic degrees of freedom refer to atomic-level translations of the two grains at the boundary. These translations, both parallel and perpendicular to the boundary surface, act so as to minimise the grain boundary free energy. They are not measurable using standard techniques for investigation of grain boundary parameters (e.g. EBSD), and will not be considered further here.

2.1.3 The interface-plane scheme

The interface-plane scheme is a useful descriptor of grain boundary geometry since it demonstrates vividly the role of the boundary plane.[19] This is because the emphasis is on the two crystallographic surfaces which form the grain boundary, denoted by the normal to the boundary surface in the

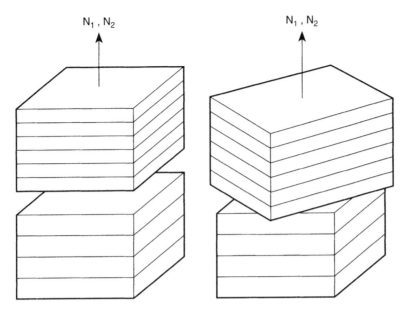

Figure 2.2 Schematic illustration of plane stacks on either side of a grain boundary with the boundary plane normal denoted *N1* and *N2* in each grain respectively. In (a) the twist angle between the plane stacks about their normal is zero and in (b) the twist angle is greater than zero.[49]

reference system of each interfacing grain. These two directions require two degrees of freedom each. In the general case the plane stacks from each grain which join at a boundary are twisted with respect to each other about the normal to the boundary plane, as illustrated in Figure 2.2. The twist angle ϕ, between the two interfacing plane stacks about their common normal, provides the fifth degree of freedom and thus the boundary is described completely. Application of the interface-plane scheme to CSL geometry is described in Section 2.2.5 in this chapter.

2.1.4 The misorientation scheme

The misorientation scheme is by far the most common method of describing grain boundary geometry.[17] It starts from the standpoint of the relative rotation between the orientations of two neighbouring lattices. The relative rotation is described concisely by an angle of rotation (misorientation) θ through which one lattice is rotated, about an axis of rotation (misorientation) UVW which is common to both lattices. The operation of this 'angle/ axis pair', as it is often called, is shown in Figure 2.3. Three degrees of freedom are required to specify the misorientation. The misorientation is frequently used alone to describe grain boundary geometry and can be

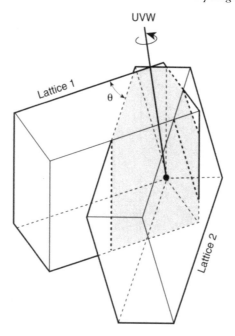

Figure 2.3   Schematic illustration of a misorientation between two lattices brought about by their notional interpenetration. The misorientation is defined by the rotation angle θ about the common axis UVW.[17]

readily related to CSL geometry as shown in Section 2.2. The remaining two degrees of freedom correspond to the crystallographic indices of the grain boundary plane with reference to *one* of the interfacing lattices. Thus the grain boundary plane is specified separately to the misorientation.

Tilt and twist components are important concepts in grain boundary geometry. Pure tilt and twist boundaries have the misorientation axis perpendicular and parallel to the grain boundary normal respectively. Although many boundaries in bicrystals are fabricated so as to be pure tilt or twist, most boundaries in polycrystals have mixed tilt and twist character. Tilt and twist components are especially significant when specifying boundary planes in CSL systems, as discussed in Section 2.2.5.

In order to perform any mathematical manipulations, e.g. calculating how much a particular misorientation deviates from a CSL, the misorientation must be expressed not as an angle/axis pair but as a 3 × 3 orthonormal misorientation matrix **M**, analogous to an orientation matrix which specifies the texture of a single grain. If two interfacing grains are denoted grain 1 and grain 2, the matrix columns are the direction cosines of the crystal axes of grain 2 referred to the coordinate system of grain 1, the reference grain. The nine numbers of the matrix overdetermine the misorientation, since it embodies only three independent variables. The angle/axis pair is obtained from the matrix as follows:

$$\cos \theta = (a_{11} + a_{22} + a_{33})/2 \qquad (2.1a)$$

$$U:V:W = a_{32} - a_{23} : a_{13} - a_{31} : a_{21} - a_{12} \qquad (2.1b)$$

If $\theta = 180°$, then UVW is given by:

$$U:V:W = (a_{11} + 1)^{\frac{1}{2}} : (a_{22} + 1)^{\frac{1}{2}} : (a_{33} + 1)^{\frac{1}{2}} \qquad (2.1c)$$

where a_{ij} (i = 1 to 3, j = 1 to 3) are the elements of the misorientation matrix. An example of a misorientation matrix, which has an axis/angle pair of 38.2°/111, is

$$\begin{matrix} .857 & -.213 & .275 \\ .275 & .857 & -.213 \\ -.213 & .275 & .857 \end{matrix}$$

When a misorientation is generated the axes of the first (reference) grain are taken to be fixed with respect to the axes of the second grain. The symmetry of the crystal system dictates that the axes of the second grain can be chosen in more than one way (except for the triclinic system). For the cubic system this multiplicity is 24; in other words there are 24 different – but equivalent – ways in which the misorientation matrix can be expressed. These symmetry-related solutions **M'** are given by:

$$\mathbf{M'} = \mathbf{T}_i \mathbf{M} \qquad (2.2)$$

where i = 1, 2 ... .24. The **T**-matrices describe the symmetry of the cubic system.[23] The 24 matrices (which contain the same elements with rows and signs interchanged) so generated give rise to 24 angle/axis pairs. Although these angle/axis pairs are entirely equivalent, it is conventional to quote the lowest angle solution when describing a misorientation. Thus, low angle boundaries (defined here as <15° misorientation) are immediately obvious.

2.2 COINCIDENCE SITE LATTICE GEOMETRY

2.2.1 Introduction

This section on coincidence site lattice geometry will focus firstly on the evolution of the CSL as a useful analysis tool and secondly on summarising the fundamental aspects of CSLs necessary to an understanding of the following chapters. More rigorous treatments of the mathematics pertaining to the CSL can be found elsewhere.[20,24]

2.2.2 Historical development

The inception of the CSL is attributed to an article reported in 1926 where it was shown that if two lattices are related by a rotation of 180° about the normal to a rational plane, then a fraction of lattice points from both lattices will coincide.[25] Later, in 1949, it was demonstrated experimentally that the well-defined relationship between many secondary and primary recrystallised grains in copper could be described by a few definite angles of rotation about low index axes.[1] These were +38°/111, −22°/111 and +19°/100. Today these would be referred to as close to $\Sigma 7$, $\Sigma 21a$ and $\Sigma 37a$ respectively. The authors first represented and explored the nature of the orientation relationships by drawing two layers of 111 planes, *A* and *B*, mutually rotated by 38° (or by −22° if the rotation is considered to be in the opposite sense) about an arbitrary point in the net as shown in Figure 2.4. It

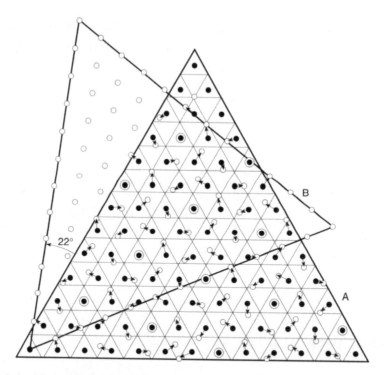

Figure 2.4 'Coincidence plot' of two 111 planes where net B is rotated by 22° with respect to net A (or 38° in the opposite direction). 1 in 7 of the lattice sites from both nets coincide, i.e. this represents a $\Sigma 7$ CSL.[1]

was observed that for this case 1 in 7 atoms coincided when the layers were rotated by 38°. These were referred to as 'coincidence plots'. The authors also observed that the two relationships 38°/1 1 1 and −22°/1 1 1 are twin related. This point will be returned to in the next subsection.

A coincidence plot, as shown in Figure 2.4, is the most visual way to illustrate the concept of a CSL, at least in two dimensions. For certain discrete misorientations between interpenetrating lattices a proportion of lattice sites will coincide, forming a periodic sublattice in three dimensions. The parameter Σ is the volume ratio of the unit cell of the CSL to that of the crystal lattice, or equivalently Σ is described as the reciprocal density of coinciding sites.[24] The actual grain boundary plane is realised as a plane running through the CSL. On average, the plane will intercept 1 in Σ atoms. It follows that the most densely packed plane in a CSL will be composed totally of coincidence sites and that some planes will contain no coincidence sites at all.[24]

The periodic structure of CSL boundaries was fully recognised in 1966 when it was first demonstrated experimentally using field ion microscopy.[27] 'Good fit' regions were observed where the boundary followed the most densely packed plane, and a stepped structure was generated where the boundary was not parallel to the densely packed plane. These observations led to the idea that deviations from the exact CSL at a boundary were accommodated by dislocation arrays, i.e. the boundary could be considered as a CSL with a dislocation sub-boundary superimposed on it.[28]

Real interest in CSLs was engendered when it was demonstrated that properties of CSL boundaries could be radically different from those of general boundaries. One of the earliest, and most quoted, studies was in 1959 and showed that the migration rate of general boundaries in zone-refined lead was dramatically reduced by very low concentrations of tin, whereas the migration rate of CSLs was hardly affected for the same concentrations.[2] These data are shown in Figure 2.5.

Two key concepts then emerged in the application of the CSL theory to real grain boundary behaviour. Firstly, a quantitative description of the observation that most boundaries in polycrystals will be 'off-coincidence' and secondly development of the central idea that the CSL is only valid physically at the boundary itself. These led in 1968 to the 'coincidence-ledge-dislocation' theory which showed that descriptions of the grain boundary plane can equivalently be by a two-dimensional array of coincidence atoms in the boundary plane, atomic ledges in the boundary or dense dislocation arrays.[22] This introduced the concept of *boundary coincidence* rather than lattice coincidence.

The CSL model formed the starting point for a more generalised coincidence model known as the *O-lattice*.[28] Whereas the CSL changes discontinuously, the O-lattice changes smoothly between point coinci-

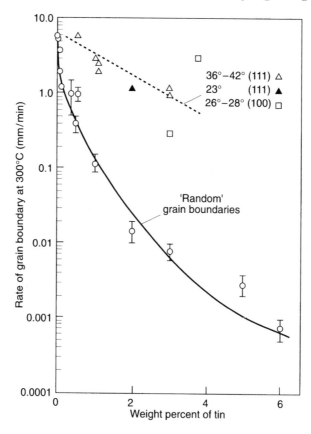

Figure 2.5 Grain boundary migration rates at 300°C in a zone-refined lead-tin alloy as a function of tin concentration. A group of CSL boundaries (near $\Sigma 7$, $\Sigma 21a$ and $\Sigma 17a$) exhibit much faster migration rates than non-CSL boundaries.[2]

dences, whether or not they are lattice points. The O-lattice was the most elegant model for describing grain boundary geometry, and provided a formal dislocation model of an interface.

Dislocations which conserved the periodicity of the CSL in the boundary plane were later termed 'secondary intrinsic grain boundary dislocations' to distinguish them from 'primary intrinsic grain boundary dislocations' which conserve the structure of the parent lattice, i.e. comprised a low-angle boundary. The Burgers vector of the secondary dislocations comprised a new lattice, the Displacement Shift Complete, DSC, lattice, otherwise known as the O_2 lattice. A description of this lattice was published in 1974.[24] These dislocation arrays in CSLs were also verified experimentally from 1969 onwards by observation in the TEM. The first examples were in a near-$\Sigma 5$ and a near-$\Sigma 3$ boundary in aluminium[29] and a near 17b boundary in an Fe-Mn alloy.[31]

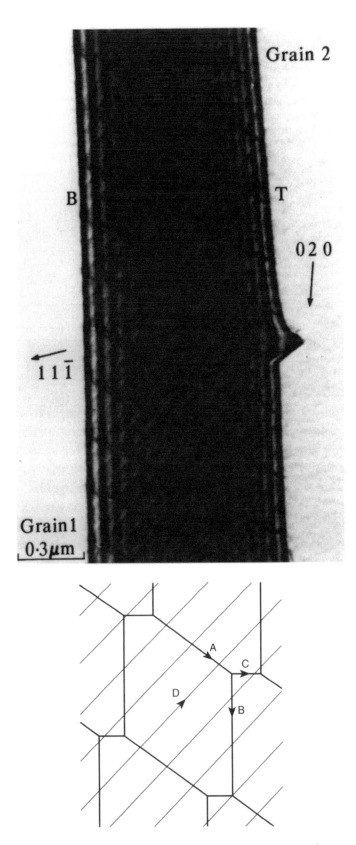

Figure 2.6 (a) Secondary intrinsic grain boundary dislocations in a $\Sigma 27a$ boundary in Cu-6at%Si alloy. The three independent arrays are labelled A, B, D in (b). C is a reaction product from the segments A and B.[146] (Courtesy of C.T. Forwood.)

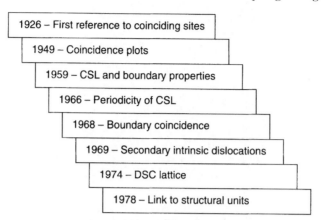

Figure 2.7   Chronology of notable stages in the development of the CSL model and its application to the study of grain boundaries.

Figure 2.6 shows an example of secondary intrinsic dislocations in a Σ27a boundary in a Cu–6at%Si alloy.[31]

An extension to the idea of boundary coincidence is to consider the density of coincidence sites actually in the boundary plane, the planar coincience site density, PCSD. Although this parameter is described in the literature in 1964,[27] it receives more attention later.[32] Essentially, for every plane where the PCSD = 1 (i.e. every site in the boundary is a coincidence site) there are (Σ-1) planes where the PCSD = 0. These ideas introduce the importance of the grain boundary plane in coincidence site lattice theory.

In the late 1970s an alternative view of grain boundary structure was proposed whereby any boundary may be viewed as a combination of poly-hedra or 'structural units'.[33,34] The parallelism between the structural unit model and the CSL model is that for certain geometries boundaries are composed of a simple combination of periodically arranged units. The lowest energy boundaries comprise only one structural unit and these have been termed *favoured*.[35] The periodicity of the units has been found to correlate with the cores of secondary grain boundary dislocations thus providing a link between the atomistic and geometrical descriptions of grain boundaries.[36]

This subsection is a short historical overview giving the early development of the CSL and its application to grain boundaries. In summary, the important milestones are displayed on Figure 2.7. Further details can be found in the papers cited above, and in recent reviews[4,9,10] which include many examples of the relationships between CSL geometry and properties.

2.2.3 Formalism of the coincidence site lattice

Generation of Σ-values. A generating function can be used to obtain values of

Table 2.1   Disorientation angle/axes for CSLs with Σ-values up to 35

Σ	θ°	UVW	Σ	θ°	UVW
3	60	1 1 1	23	40.5	3 1 1
5	36.9	1 0 0	25a	16.3	1 0 0
7	38.2	1 1 1	25b	51.7	3 3 1
9	38.9	1 1 0	27a	31.6	1 1 0
11	50.5	1 1 0	27b	35.4	2 1 0
13a	22.6	1 0 0	29a	46.3	1 0 0
13b	27.8	1 1 1	29b	46.4	2 2 1
15	48.2	2 1 0	31a	17.9	1 1 1
17a	28.1	1 0 0	31b	52.2	2 1 1
17b	61.9	2 2 1	33a	20.1	1 1 0
19a	26.5	1 1 0	33b	33.6	3 1 1
19b	46.8	1 1 1	33c	59.0	1 1 0
21a	21.8	1 1 1	35a	34.0	2 1 1
21b	44.4	2 1 1	35b	43.2	3 3 1

Σ and accompanying values of the angle of misorientation, θ, and axis of misorientation, UVW, in the cubic system.[26] If UVW is chosen, then θ and Σ are given by:

$$\Sigma = x^2 + Ny^2 \qquad (2.3a)$$

$$\tan(\theta/2) = yN^{\frac{1}{2}}x \qquad (2.3b)$$

where $N = U^2 + V^2 + W^2$ and x, y are integers ≥0. In practice only CSLs having a relatively short periodicity, i.e. low Σ, are of interest and these are shown in Table 2.1. Usually it is sufficient to know only the lowest angle solution – the '*disorientation*' – and a misorientation is usually quoted in this form. The other symmetry-related solutions can be readily calculated if necessary using equation 2.2 given in section 2.1.4. More detailed tables of CSLs, including all 24 symmetry-related solutions, are provided else-where.[17,37]

CSLs at triple junctions. There are geometrical rules governing the relation-ship between three CSLs which meet at a triple junction.[26] These are:

• They share a common misorientation axis;
• The sum of two of the misorientation angles gives the third;
• The product or quotient of two of the Σ-values gives the third.

In the previous subsection it was remarked that in the earliest reported observation of CSLs the authors noted that the coincidence plots of $+38°/111$ and $-22°/111$ were twin related. Using the rules given above, this can be expressed as:

$$60°/111 - 21.79°/111 = 38.21°/111 \qquad (2.4)$$

$$\text{i.e. } \Sigma3 - \Sigma21a = \Sigma7.$$

The CSLs combination which has the most important practical implications in terms of grain boundary engineering concerns interactions of the $\Sigma3''$ family. These can include either twinning or dissociation[38] and the most common example is the meeting of two $\Sigma3$ boundaries:

$$70.53°/110 + 70.53°/110 = 141.06°/110 \qquad (2.5)$$

$$\text{i.e. } \Sigma3 + \Sigma3 = \Sigma9.$$

Proportions of CSLs expected. The probability of CSL occurrence has been assessed using both a random number generator and a probability density function.[39] 11% CSLs with $\Sigma \leq 25$ are expected for randomly oriented, unconnected grain pairs. This figure includes 2% low angle boundaries. Computer simulations of CSL proportions are discussed further in Section 4.8.

CSLs in non-cubic systems. Although it is not an explicit topic in this book, it should be mentioned that the CSL formalism can be extended to non-cubic materials.[40,41,42] The determination of CSLs in non-cubic systems depends critically on the axial ratios of the lattice parameters, and so is more complex than for the cubic case. The concept of 'near coincidence' or 'constrained coincidence' has been developed to extend the range of allowable CSLs.[43] Some particular areas of interest include superconducting materials, ceramics and interphase boundaries.

2.2.4 Deviations from exact coincidence

The requirement for CSL analysis in polycrystals is to identify all misorientations which are 'close to' CSLs (up to a predetermined Σ value) in terms of an angular deviation. The maximum deviation, v_m, equates in physical terms to the maximum conservation of the CSL by secondary dislocations. There is no definitive criterion which stipulates v_m, although there are physically based principles which have been used to propose guidelines for it. The experimentally measured deviation is denoted v, and deviations from exact CSL can be described by a relative deviation, v/v_m.

The structural analogies between low angle boundaries and CSLs led to the adoption for the CSL case of the following well-known Read–Shockley relationship which links the dislocation density d in the boundary with the Burgers vector b and the angular misorientation δ:[44]

$$\delta = b/d \text{ (low angle)} \tag{2.6a}$$

$$v_m = b/d \text{ (CSL)} \tag{2.6b}$$

Thus, v_m corresponds to the highest density of dislocations possible in the boundary. The density of dislocations which can be accommodated in a CSL is related to its periodicity, i.e. Σ.[45] The variation of v_m with Σ is usually taken to be as $\Sigma^{-\frac{1}{2}}$, the so-called 'Brandon criterion', which corresponds to a relationship based on periodicity alone.[46] Hence

$$v_m = v_0 \Sigma^{-\frac{1}{2}} \tag{2.7}$$

where v_0 is a proportionality constant based on the angular limit for a low angle boundary, 15°. If $\Sigma = 1$ is substituted into equation 2.7, v_m is 15°, which is the low angle boundary limit. Hence a low angle boundary can be described as $\Sigma 1$.

The Brandon criterion has been adopted almost universally for deciding v_m. Other criteria have proposed that v_m should be proportional to $\Sigma^{-2/3}$, or Σ^{-1} (for twist boundaries) or $\Sigma^{-5/6}$.[17] The last of these is supported by correlation with special properties and also observation of secondary dislocations, as shown in Figure 2.8. The geometrical basis for the $\Sigma^{-5/6}$ dependency is that d is proportional to the mean edge of the CSL cell, $\Sigma^{1/3}$, and b is

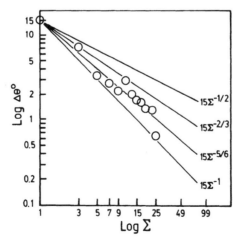

Figure 2.8 Maximum deviation angle $\Delta\theta$ from CSLs in 99.999% nickel which display selective immunity to intergranular corrosion.[65]

proportional to $\Sigma^{-\frac{1}{2}}$. Substitution of these values in equation 2.6b gives $\Sigma^{-5/6}$.[47]

Although arguably the Brandon criterion may not be the most accurate descriptor on a physical basis, for reasons mainly of consistency between investigations it is used almost universally as the CSL deviation cut-off.

2.2.5 Tilt and twist components in the coincidence site lattice

As described in Section 2.1.3, the interface-plane scheme is a means of describing the geometry of both misorientations and grain boundary planes. It is even more powerful for CSL boundaries, where it is especially relevant to identify tilt and twist components and low index boundary planes, since these ultimately control grain boundary properties (see section 5.1). There are four boundary types, categorised according to the interface-plane scheme: twist grain boundaries (TWGBs), symmetrical tilt grain boundaries (STGBs), asymmetrical tilt grain boundaries (ATGBs) and boundaries which are none of these but are nonetheless characterised by low-index planes. The way in which each of these is recognised in the interface-plane scheme will now be described.

In the scheme **N1** and **N2** are the directions of the grain boundary plane normal referenced to grain 1 and 2 respectively, either side of the boundary. ϕ is the twist angle between plane stacks from each interfacing grain.[48]

Twist boundaries. A TWGB has the twist axis perpendicular to the axis of misorientation, and is characterised by the condition **N1** = **N2** and $\phi > 0$. All the TWGBs for a particular CSL are given by *UVW* for each of the 24 symmetry related solutions.

Tilt boundaries. A tilt boundary has the tilt axis parallel to the axis of misorientation. A STGB has planes from the same family (i.e. the same form of Miller indices) on either side of the boundary whereas an ATGB has different plane types interfacing the boundary. Both of these tilts are recognised in the interface plane scheme by $\phi = 0$.

If $\phi = 0$ and **N1, N2** are different but commensurate, the boundary is an ATGB. A commensurate boundary arises when the ratio of the planar grain boundary unit cells is an integer G. The planar grain boundary unit cell area is proportional to the separation of planes parallel to the boundary. G is therefore given by the square root of the ratio of the sum of the squares of the Miller indices. For example, $\{751\}$ and $\{111\}$ are commensurate planes because

$$G = ((7^2 + 5^2 + 1^2)/(1^2 + 1^2 + 1^2))^{\frac{1}{2}} = 5$$

i.e. an integer. These two planes constitute an ATGB in the $\Sigma 5$ system.

A CSL can give rise to two STGBs at the most but many ATGBs. This is one reason why ATGBs have been found to be ubiquitous in polycrystals (see Section 5.1). To assist in identifying ATGBs, tables have been published giving lists of commensurate planes for each Σ system.[17]

Low index planes. There is some evidence that boundary planes which are near densely packed lattice planes are important in polycrystals;[49,50] this will be discussed in detail in Section 5.1. These planes are not related to a low-Σ CSL, i.e. the grain boundary is not periodic. On the other hand they may be only a few degrees from an ATGB configuration, such that the mismatch could be accommodated by dislocation arrays. Table 2.2 lists some ATGBs which are near low-index combinations, that is, are within 5° of a 111, 100 or 110 plane in one grain, and a low-index plane in the other grain.

Table 2.2 Asymmetrical tilt boundaries having one plane within 15° of
111, 100 or 110

Plane close to 111			Plane close to 100		
Σ	Planes	Deviation from plane	Σ	Planes	Deviation from plane
7	210/10,98	1.5	11	111/19,11	4.3
9	110/877	3.7	17	210/38,10	1.5
11	100/766	4.3	17	110/24,11	3.4
11	221/20,20,17	4.3			
11	311/23,21,19	4.5	Plane close to 110		
17	211/25,25,22	3.4			
17	211/26,23,23	3.4	9	111/11,11,1	3.7
19	310/36,35,33	2.1	11	211/19,19,2	4.3
19	310/39,35,35	3.0	13	210/21,20,2	4.2
21	221/38,37,34	2.7	13	210/22,19,0	4.2
23	210/31,30,28	2.4	13	211/23,22,1	2.2
25	211/37,35,34	2.0	13	221/29,26,2	4.3
27	110/23,23,20	3.7	13	310/31,27,0	3.9
			15	310/35,32,1	2.8
			17	100/12,12,1	3.4
			17	221/38,34,1	3.7
			17	21/38,34,1	3.4
			17	310/39,37,0	1.5
			27	111/35,31,0	3.7

2.3 GRAIN BOUNDARY PROPERTIES

2.3.1 Introduction

The reason for applying the CSL system in a grain boundary engineering context is as a convenient link between grain boundary *geometry* and *properties*. Before we can examine the experimental evidence for this usage, it is necessary to specify and characterise the actual 'grain boundary properties' themselves, which is the topic of this section.

2.3.2 Fundamental properties

Grain boundaries have different properties from the lattice because, at a local level, the average atomic volume in a grain boundary is greater than in the grain interior.[6] In other words there is *free volume* at a grain boundary, as illustrated in Figure 2.9. Excess free volume and lowered atomic coordination are the fundamental intrinsic properties of a grain boundary – that is, what differentiates it from the lattice. Furthermore, it is the hallmark of special boundaries that they have lower than average free volume and this in turn is what distinguishes them from general boundaries.

Boundary free volume has been studied by hard sphere modelling,[34] computer modelling using the embedded-atom-method[51] and experimental obervations using HREM.[50] The end result of these investigations is that a

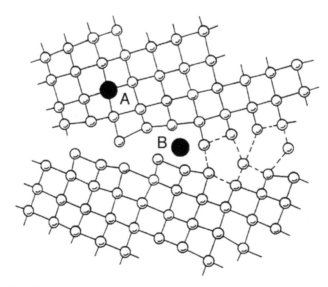

Figure 2.9   Simple schematic illustration of the lattice sites in the vicinity of a grain boundary. Solute atoms may occupy a lattice site (A) or reside within the boundary core (B). (Courtesy H. Gleiter.)

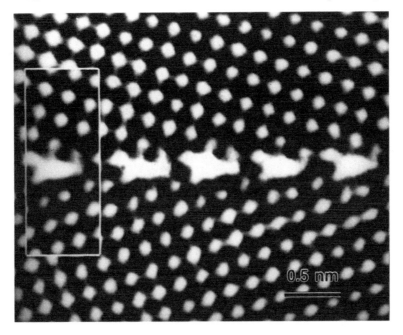

Figure 2.10 Atomic resolution micrograph of a symmetrical tilt Σ5 CSL on (310) in NiO[129]. (Courtesy of K.L. Merkle.)

variety of low-Σ CSLs with the boundary plane in the symmetrical or asymmetrical tilt position have low free volume therefore are truly special boundaries. It is also an essential feature of these boundaries that they are composed of periodic polyhedral groups of atoms as mentioned in section 2.2.2. Figure 2.10 is a HREM image of a Σ5 (310) STGB in NiO illustrating the atomic positions at the boundary.

Since grain boundaries are essentially defects, they have an energy associated with their non-equilibrium structure. Computer modelling techniques combined with HREM have demonstrated that the boundary excess free volume correlates monotonically, but not always linearly, with the boundary energy.[6,51] In the literature CSLs are often referred to as 'low energy' boundaries, even though the link with low energy has only been demonstrated for tilt and some twist cases. Where a CSL boundary does not have this configuration its energy will be higher.

Actual values of grain boundary energies range from less than 200 mJm^2 for some tilt and twist boundaries to more than 900 mJm^2 for a completely disordered, general boundary. Figure 2.11 shows the range of energies for some simulated tilt and twist CSLs, correlated with their free volumes.[51] The energy of most of these boundaries lies between 300–700 mJm^2, except for

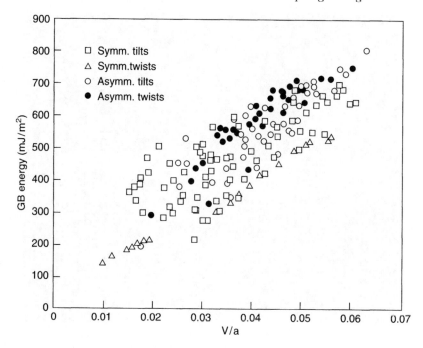

Figure 2.11   Calculated correlation (for fcc metals) between grain boundary energy and volume expansion per unit boundary area (V/a).[51]

the special case of the 1 1 1 symmetrical twin boundary whose energy is an order of magnitude less than that for other boundaries. It is also significant to note that the energies of some asymmetrical boundaries is lower than those which are symmetrical. These points are discussed further in Section 5.1.

Experimentally obtained grain boundary energies – that is, from real materials rather than simulations – present some ambiguities. Early classic work involving fcc tilt boundaries misoriented on 1 1 0 demonstrated deep energy cusps corresponding to $\Sigma 3$ and 11 STGBs on the 1 1 1 and 3 1 1 planes respectively as shown on Figure 2.12.[52] This work on 1 1 0 tilt boundaries has been verified several times since, with additional shallow cusps identified for $\Sigma 9$ and $\Sigma 27$.[19] However, cusps for boundaries such as $\Sigma 7$ have not been convincingly identified, although they have been inferred from rotating sphere experiments.[3] These latter experiments involve allowing energy minimisation to take place between freely rotating metal spheres and a substrate, and identifying the misorientation peaks which result.

Other early work showed that $\Sigma 25$, 13, 17 and 5, which were all 1 0 0 twist boundaries, occurred in CdO 'smoke'.[53] The CdO crystals, with surfaces parallel to 1 0 0, were obtained by burning Cd rods, collecting the oxide particles which formed and measuring the misorientation distribution

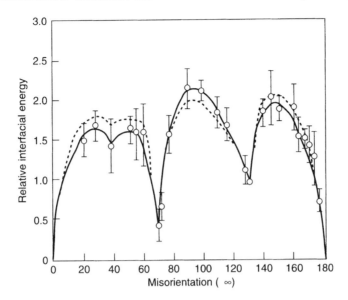

Figure 2.12   Measured relative energies of [1 1 0] tilt boundaries in aluminium as a function of the misorientation angle.[52]

which resulted. The distribution showed very well defined peaks for CSLs misoriented on 1 0 0, thus inferring low energy for these interfaces. However, in a parallel experiment where MgO was substituted for CdO, the peak for $\Sigma 17$ was missing from the spectrum of CSLs on 1 0 0. This anomaly demonstrates the limitations of a purely geometrical model to predict energy, since energy is a function of effects related to the chemical nature and electronic interactions of the species in addition to geometry.

Other grain boundary properties, which have different values to those in the lattice, arise directly from the boundary free volume. These are intrinsic resistivity,[54] segregation and diffusivity, including the ability of the boundary to act as a source/sink for vacancies.[55] Since these intrinsic boundary properties depend directly on free volume, low free volume boundaries will exhibit low values for these parameters, exactly as for boundary energy, which was discussed above. The intrinsic properties depend only on the structure of the *core region* of the boundary, i.e. up to about 5 nm either side of the boundary plane. Figure 2.13 shows a comparison between the lattice and grain boundary diffusion coefficient to illustrate that the differences in value between the lattice and grain boundary case can span orders of magnitude.

Another important diffusion-related grain boundary property is mobility, i.e. the ability of the boundary to migrate or slide under the action of a driving force. A simplified expression for intrinsic mobility, *m*, shows that it depends directly on the boundary diffusivity:[56]

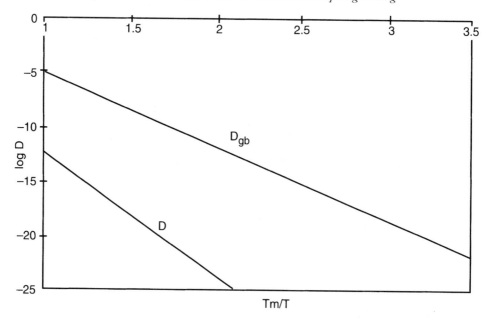

Figure 2.13   Comparison between the grain boundary diffusion coefficient, D_{gb} and lattice diffusion coefficient, D for a range of homologous temperatures T_m/T.[17]

$$m = (a^2/kT) \, D_{gb} \qquad (2.8)$$

where a is the lattice parameter, D_{gb} is the diffusivity in the grain boundary and k, T have their usual meanings.

In summary, grain boundary property parameters are:

• Energy
• Diffusivity
• Segregation
• Mobility
• Resistivity.

These properties all depend directly on the level of free volume at the boundary. Some tilt and twist boundaries, i.e. CSLs having particular boundary planes, are characterised by a lower than average free volume and therefore are associated with property values which differ from the average.

2.3.3 Grain boundary phenomena

The grain boundary properties listed in the previous subsection give rise to measureable grain boundary 'phenomena'. These include:

• *Migration and sliding.* Movement of a boundary parallel to or perpendicular to its plane, i.e. grain boundary sliding and migration respectively, are functions of grain boundary mobility and, in turn, diffusivity (Equation 2.8). In impure materials the rate of grain boundary movement is also influenced by the presence of solutes in the boundary, giving rise to 'solute drag'. Grain boundary migration underpins phenomena of major importance in materials processing, namely recrystallisation and grain growth (including anomalous grain growth) as well as other phenomena such as 'diffusion induced grain boundary migration', DIGM, etc. Grain boundary sliding occurs at high temperatures and underpins creep and superplasticity.

• *Precipitation.* Precipitation of second phases at the boundary depends on segregation of solute species to the boundary plane and, in some cases such as discontinuous precipitation, boundary migration. The kinetics of precipitation at grain boundaries differs from that in the lattice.

• *Intergranular degradation.* A range of intergranular degradation processes e.g. cavitation, fracture, corrosion and embrittlement can occur in materials under the action of stress and/or aggresive environments. These are of course deleterious for components in service and are a major cause of failure. Such phenomena are related to diffusivity, segregation and in certain circumstances mobility.

• *Energy minimisation.* Various grain boundary equilibration processes, for example annealing twinning, grain rotation, grain boundary recovery and grain growth can operate under the influence of a thermally activated driving force with the result of lowering the overall energy of the system. The first three achieve this by reducing the energy of individual boundaries whereas grain growth acts by reducing the total grain boundary area of the system.

This grain boundary phenomena categorisation does not imply that each is mutually exclusive. There is an underlying common dependency on boundary structure which links various manifestations of boundary behaviour. For example low energy boundaries are also those which are resistant to intergranular degradation. All the phenomena listed above show sharp discontinuities in values for special boundaries, which have formed the subject of many investigations. These investigations have frequently been classified and analysed via the CSL framework, and the cases which relate to polycrystals are comprehensively reviewed in Chapter 4.

These phenomena are more self-evident and easier to assess than the underlying intrinsic properties, although the interpretation of such observations and measurements is often more complex. Interpretation of grain boundary migration rates is a case in point. It has long been known that small amounts of solute which segregate to the boundary can change the migration rate by orders of magnitude in general boundaries, whereas certain CSL boundaries are largely unaffected.[2] The steady-state migration rate V is a function of both the intrinsic mobility *M* and the acting driving force *P*:

$$V = MP^n \qquad\qquad (2.9)$$

where $n \approx 1$ except for special circumstances.[57] It might be inferred, therefore, that solutes (impurities) modify the mobility, but in fact the effect of solutes is on the *driving force* via solute drag rather than on the intrinsic mobility, i.e. on *P* rather than on *M* in equation 2.9.[56] Reported correlations between boundary misorientation and mobility may therefore be a segregation effect. It is clearly vital to understand the nature of these interactions in order to interpret correctly the experimental results.

There are other complications which impact upon interpretation of the link between boundary geometry, as designated by the CSL, and properties. One of these is the effect of lattice dislocations which both enter the boundary and also cause a long range stress field. These are known as 'extrinsic grain boundary dislocations', EGBDs. Figure 2.14 shows a lattice dislocation entering a grain boundary and being absorbed by it. The CSL model does not take account of the effect of these defects, since the model is purely geometric in nature. Some studies have shown that at elevated temperatures EGBDs are incorporated more rapidly into general high angle boundaries than into CSLs, in accordance with the higher free volume of the former.[58] Other investigations do not demonstrate a firm link between absorption of EGBDs and CSLs.[59] The 'disappearance' of extrinsic dislocations in boundaries and its relationship with boundary diffusivity is discussed further in Section 4.9.2.

EGBDs strongly modify grain boundary behaviour. Some examples are:

• It is considered that one of the mechanisms for grain boundary sliding is the movement of extrinsic (i.e. from the lattice) dislocations in the boundary plane.[60]

• Segregation is affected by EGBDs since they can act as concentration sites for solutes, leading to precipitation on dislocations or preferential diffusion.[61]

• The role of EGBDs on grain boundary migration is clearly highlighted by the phenomenon of 'strain-induced grain growth', whereby strains of the order of 2% induce anomalous grain growth, i.e. the rapid migration of a few boundaries.[62]

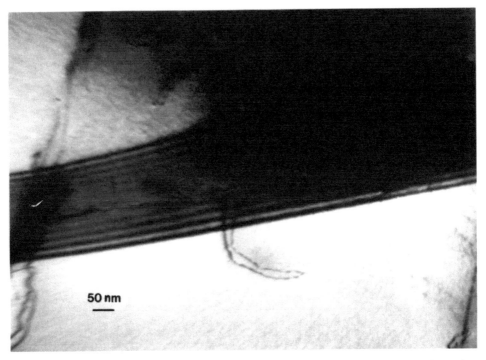

Figure 2.14 Lattice dislocations entering and being absorbed into a random high-angle grain boundary in a nickel-based superalloy.

2.3.4 Factors which affect grain boundary behaviour

To summarise the discussion in the preceding parts of this Chapter, for a particular material grain boundary response and behaviour is a consequence of the following combination:

• Grain boundary geometry;
• Grain boundary solutes;
• Extrinsic grain boundary structure.

Grain boundary geometry refers to the basic building-blocks of the boundary which, for the case of a low-Σ CSL with a rational boundary plane, has a periodic structure. For a superpure, well annealed metal boundary behaviour depends principally on the geometrical component. However most materials do not fall into this category and, as demonstrated in the previous subsection, there is often a contribution from both solutes in the form of a drag effect and non-equilibrium grain boundary structure in the form of vacancies or EGBDs. In practice it is usually impossible to isolate the role which individual factors have on grain boundary behaviour, or the relative importance of each of them. It has been suggested that the properties

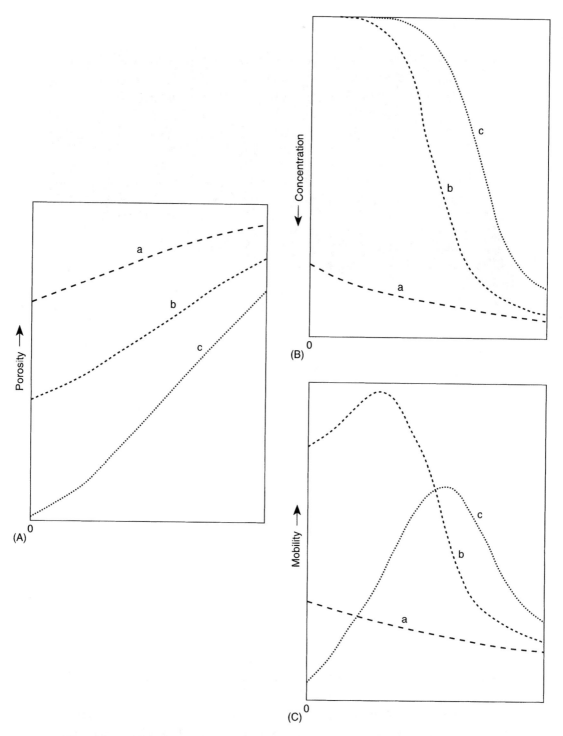

Figure 2.15 Schematic plots of (A) grain boundary porosity (i.e. free volume), (B) concentration of segregant and (C) the sum of A and B, as a function of deviation from the exact misorientation, $\Delta\theta$. General high angle boundaries, medium density and high density CSLs are denoted by curves a, b and c respectively.[63,64]

of boundaries depend more on the defect structure (solutes, vacancies, EGBDs) than on the ideal structure, exactly analogous to the case for the lattice where dislocations are responsible for plasticity rather than the equilibrium structure of the lattice.[5]

Attempts have been made to rationalise the synergy between the factors which affect grain boundary behaviour. For example, some quite early grain boundary work proposed that the effective grain boundary mobility is a sensitive function of both porosity (i.e. free volume) and concentration (segregation at the boundary).[63] Schematic curves representing these parameters as a function of deviation from the exact CSL are shown in Figure 2.15. The three curves represent the case for general boundaries (i.e. high Σ), medium-Σ CSLs and low-Σ CSLs (e.g. $\Sigma 3$), curves A, B, C respectively. Summation of the porosity and concentration plots gives the effective mobility. This shows that whereas general boundaries are unaffected by deviation, CSLs have a maximum effective mobility which is displaced from the exact CSL configuration. Although this is in general agreement with observation – for example, coherent twins, which have the exact $\Sigma 3$ configuration, are immobile[64] – it does not account fully for the solute effect on driving force nor the role of different grain boundary planes.

In practice, it is common in the literature for all the boundary properties, phenomena and behaviour which have been described in this subsection to be known generically as 'grain boundary properties'. To recap, these intergranular attributes are:

• Cavitation
• Corrosion
• Diffusivity
• Embrittlement
• Energy
• Fracture
• Migration
• Mobility
• Precipitation
• Resistivity
• Segregation
• Sliding

Several comprehensive reviews document correlation between all these properties and low-Σ CSLs, mostly for bicrystals.[5,9,10,65,66]

3. From Bicrystals to Polycrystals

3.1 INTRODUCTION

Much of the work which underpins the current knowledge base on the relationship between CSL geometry and grain boundary properties has been accrued from experiments on bicrystals. The advantage of such experiments is that the crystallography and chemistry of the grain boundary can be carefully controlled, and the bicrystal can be prepared as a symmetrical tilt, asymmetrical tilt or twist boundary with the results interpreted accordingly. High resolution images, coupled with computer simulation, have helped to elucidate the structure of such boundaries (e.g. Figure 2.10).

Several reviews provide extensive coverage of the data base concerning bicrystal property/geometry relationships[9, 10, 65] which will not be reiterated here since the present work is concerned distinctly with polycrystals. The overall message from the bicrystal reviews is that there are many examples which show that low-Σ boundaries in bicrystals exhibit each of the special properties mentioned in the previous Chapter.

Perhaps one of the most convincing demonstrations that CSLs represent a preferred subset of grain boundary geometry is the 'sphere-on-plate' or crystal rotation experiments. Such experiments involved large numbers of bicrystal pairs, and were among those which laid a foundation for stimulating progress in applying the CSL model to polycrystals.[3] Essentially the sphere-on-plate is a sintering study. When a crystal is allowed complete rotational freedom with respect to a second, fixed crystal with which it is in contact, thermal activation allows the first crystal to rotate into an equilibium position, i.e. a local energy minimum, with the fixed crystals. Measurements of X-ray maxima show that the spheres rotate into well-defined CSL relationships with the fixed substrate.

Crystal rotation experiments, using many thousands of crystallites on 1 0 0, 1 1 0 and 1 1 1 substrates, were conducted mainly in the 1970/80s on various materials such as MgO, CdO, copper and silver[67, 68, 69] and interpreted as clear evidence for the low energy of the CSL relationships observed. A hierarchy of energies was obtained from the amplitude of the peak for each CSL. Care had to be taken in these experiments that the energy cusps inferred were in fact associated with final rather than metastable configurations. In some cases the orientation of the interfacing planes was obtained, indicating various low energy tilt and twist configurations and indeed that the crystallographic orientation of the boundary plane is as important in determining energy as the misorientation angle and axis.[70]

In polycrystals boundaries very rarely have an exact CSL configuration. As discussed in Section 2.2.4, the deviation from an exact CSL is accommodated by secondary intrinsic dislocations. Special properties associated with the CSL structure will necessarily decay with the deviation parameter, v, which was defined in Section 2.2.4. Although it has not often been studied in a rigorous manner there is evidence that special properties are restricted to quite a narrow region about the exact CSL, providing justification for a narrower defining criterion (see Figure 2.8).[47,71]

The link between grain boundary geometry and properties has more complexities for the polycrystal than for the bicrystal case. The main aspect to be considered is the *connectivity* of a polycrystalline microstructure as illustrated in Figure 3.1.[21,72] In addition to grain boundaries, connectivity

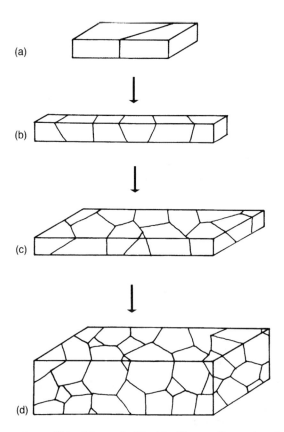

Figure 3.1   Illustration of (a) a bicrystal, (b), (c), (d) one-dimensional, two-dimensional and three-dimensional polycrystals respectively.[21]

introduces two other topological features into the three-dimensional micro-structure: *grain edges* (otherwise known as grain junctions or triple lines) and *grain corners* otherwise known as nodes. The necessary maintenance of connectivity in a three-dimensional polycrystal also restricts phenomena such as grain rotations, since changes which might lower the energy of one interface cannot occur in isolation, and other affected interfaces might not similarly benefit.[73,74]

The simplest polycrystal configuration is one-dimensional, e.g. a wire, where boundaries extend across the whole of its width as shown in Figure 3.1b. Few CSL studies have been conducted on wires, but results show that thermomechanical treatments allow a large proportion of CSLs to develop.[75] This is not suprising since the wire structure is essentially a linear series of bicrystals where none of the boundaries conjoin. Another example is the creep fracture of an Al_2O_3 film deposited on a thin copper strip where 89% of the copper boundaries were CSLs and were all resistant to fracture.[76]

A two-dimensional polycrystal (e.g. a sheet, foil or ribbon) contains boundaries which extend through the entire thickness of the sheet, plus grain junctions (Figure 3.1c).[77,78,79] Such a configuration still allows a high proportion of CSLs to develop under suitable annealing conditions, partic-ularly when accompanied by a strong texture. This is because the two-dimensional grains can rotate with relative ease. For example the high ductility of Fe-6.5%Si melt-spun ribbons was attributed to them containing 45% CSLs (including Σ1s), a strong 100 texture and a large grain size.[80]

The vast majority of materials are bulk, three-dimensional polycrystals which are different from the two-dimensional type in that they contain grain corners or nodes in addition to grain boundaries and junctions (Figure 3.1d). Here the geometry of a boundary is constrained by the orientation of several grains and processes such as energy minimisation now have to apply to the whole system rather than a single boundary as in the bicrystal case.

The frequent demonstrations of the link between special properties and CSLs in bicrystals has spawned similar experiments in polycrystals. However, from the foregoing discussion clearly it is an oversimplification to expect duplication of the results from bicrystals in polycrystals. The following points summarise the main differences between polycrystal and bicrystal experi-ments:

• Topological constraints, as described above;
• Uncertain position of the grain boundary plane;
• Uncertain defect and solute contents.

3.2 OVERVIEW OF STRATEGIES FOR POLYCRYSTAL COINCIDENCE SITE LATTICE INVESTIGATIONS

A study of all the recent literature concerning application of the CSL model to polycrystals discloses that there are several strategies in use. Often more than one approach is used in individual investigations.

- *'Bicrystal ensemble'*. The most popular rationale for CSL investigations in polycrystals is the 'bicrystal ensemble' approach. In other words, a system is treated as a collection of unrelated bicrystals for the purpose of formulating statistics. The motivation for using the CSL in this type of experiment is generally either that it is a convenient classification system (for example, it is typically employed in this way for recrystallisation experiments) or correlation is sought with a particular property. This whole group constitutes 'mainstream' CSL experiments (see Chapter 4).
- *Continuous functions*. There are several continuous functions, which are used to describe grain boundary parameters, such as the 'correlated' and 'statistical' MODF,[81] MisOrientation Distribution Function,[82] the Inter-crystalline Structure Distribution Function, ISDF[83] and the Interface Damage Function, IDF.[84] Analysis via the CSL formalism is introduced by categorisation of the peaks in these functions (see Section 4.7).
- *Computer simulation/modelling*. Computer simulation/modelling has been recognised as a powerful means of probing the predictive power of the CSL model. Frequently simulated data is compared with that obtained experimentally (see Section 4.8).[85,86,87,88,89]
- *Grain boundary plane*. The misorientation scheme approach to CSL analysis does not take account of the grain boundary plane location. A few studies have extended the measurements to a fuller description of grain boundary geometry which includes crystallographic plane analysis. This approach is described in more detail in Chapter 5.
- *Connectivity*. The key feature of a polycrystal which distinguishes it from a bicrystal is that the grain boundaries in the former constitute a *network*. This connectivity of boundaries has important consequences for phenomena such as corrosion. It is therefore not sufficient to study the behaviour of individual boundaries without giving due consideration to how they are linked to the rest of the network. Hence grain boundary studies have extended to consideration of grain junction (triple line) geometry,[21,90,91,92] and functions such as the Grain Boundary Correlation Number (GBCN),[93] which gives how many boundaries of the same type are connected to one another, have been developed (Section 5.2).

3.3 MODERN EXPERIMENTAL METHODS USED TO OBTAIN COINCIDENCE SITE LATTICE DATA

Coincidence site lattice investigations can be conducted using several different experimental methods:

- X-ray diffraction (Laue or synchrotron)
- TEM (including high voltage microscopy)
- Selected area channeling (SAC) in an SEM
- EBSD (including COM and OIM) in an SEM
- Indirect methods, i.e. etch pit analysis[94] and atomic force microscopy.[95]

The reader is referred elsewhere for more details of most of these techniques and their merits.[17,96] Figure 3.2 shows the breakdown of technique usage for all the investigations documented here, and it can be seen that SAC, EBSD and TEM are the principal methods used to characterise grain boundaries. TEM remains suprisingly popular, considering the labour intensity involved in foil preparation and data collection. Many TEM experiments involve recrystallisation, where the CSL is used as a classification system (see Section 6.1.2).

The best means for obtaining CSL data is to use a technique which gives

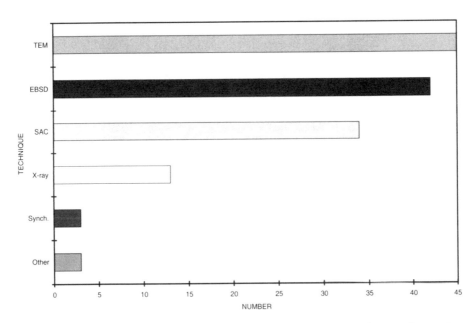

Figure 3.2   Distribution of specific experimental techniques used to conduct grain boundary (CSL) investigations on polycrystals during the last 20 years. (Synch. = synchrotron radiation.)

the orientation of individual, neighbouring grains e.g. TEM or EBSD. From these measurements the misorientation between the two grains can be obtained and then classified in the CSL system by means of the algorithms given in Chapter 2. Further measurement steps are needed to obtain the orientation of the boundary plane (see Section 5.1.4). If a technique is used which provides continuous diffraction information, i.e. averaged from many grains, a CSL classification is accessed from the misorientation between peaks in the orientation distribution.

The most notable trend is that since about 1990 EBSD has become the major technique for grain boundary studies.[13,96] The attractions of EBSD are that specimen preparation is usually minimal, the spatial resolution is approximately 0.5 μm and diffraction patterns are accurately indexed by

100.0 μm = 10 steps

Figure 3.3   Orientation imaging micrograph (OIM) featuring an intergranular crack in Inconel 750. Random high angle boundaries are delineated in black and Σ3 boundaries in grey. In the original OIM, Σ3s, Σ9s and Σ27s were coloured red, green and yellow respectively. (Courtesy D.P. Field.)

computer in real time. The latest generation of systems features automatic pattern indexing, which can be coupled with stage or beam control on the microscope. The operator can either preselect the coordinates of sampling points, or sample orientations at predetermined steps. The most sophisticated automated EBSD systems employ COM or OIM,[97] as illustrated in Figure 3.3. Here the grain boundaries have been automatically mapped from orientation changes, and CSL boundaries are shown highlighted.

4. *Coincidence Site Lattice Data Overview and Analysis*

4.1 INTRODUCTION

In order to monitor CSL trends and features, over 200 studies involving CSLs in cubic polycrystals have been collated, followed by extraction and analysis of key data. These studies have been reported in the readily available literature over approximately the last 20 years. Some data have been extracted from conference proceedings and reviews, provided that the actual results have not been published elsewhere. Thus care has been taken to avoid duplication in compiling the statistics. The following parameters, where reported, have been compiled for each investigation (listed here in alphabetical order):

- Experimental methods
- Grain size
- Material
- Maximum value of Σ
- Property studied
- Proportion of CSLs
- Proportion of Σ3s, Σ9s, Σ1s
- Sample population size
- Reference
- Results/comments.

The following grain boundary properties featured in the total investigation set:

- Cavitation or Creep
- Cracking
- Diffusivity
- Ductility
- Energy
- Fatigue
- Mobility, including recrystallisation/grain growth
- Precipitation/pinning
- Segregation

The following, while not being strictly 'properties' as defined in section

2.3, have also provided rationales for investigations and so are included in the statistics:

- Connectivity
- Orientation relationships/transformations
- Statistics of distributions
- Microstructural evolution
- Texture
- Twinning

The present Chapter contains a digest of the literature, categorised as above. A master list from which the statistics have been compiled can be found in the Appendix. Some entries in this list also appear in the numbered references. Data which are reported as continuous functions and computer simulations are included in the statistics, and the methodology for these subtopics is discussed in Sections 4.7 and 4.8 respectively.

4.2 MATERIALS

Figure 4.1 gives an overview of the materials which have been the subject of CSL investigations. The first group of materials are face-centered cubic (fcc),

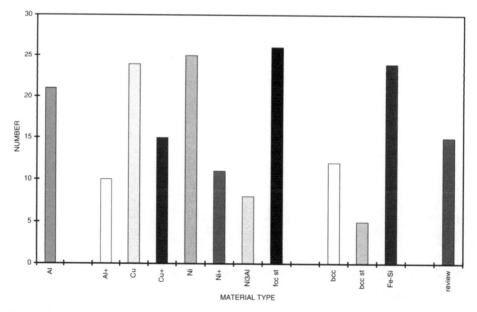

Figure 4.1   Distribution of CSL investigations on polycrystals according to material type. Al+, Cu+, Ni+ refer to alloys of these elements.

followed by a group of body-centered cubics (bcc), and a category of review articles which by definition includes several investigations, often on a range of materials. With regard to the fcc groups, the major groups investigated are pure (or nearly pure) materials aluminium, copper and nickel and fcc steels (denoted 'fcc st' on Figure 4.1). Al+, Cu+ and Ni+ signifies alloys or compounds of these elements. Ni_3Al has been placed in a separate category because there have been a relatively high number of investigations on this intermetallic. Turning to the bcc group, there have been only about one-quarter of the number of investigations performed as on fcc materials. The 'bcc' group includes iron, tungsten and molybdenum, and the 'bcc st' group includes a range of ferritic steels. A major group in the bcc category is the Fe-Si alloys.

The distribution of materials reflects both basic research and commercial interests, with the more fundamental studies of grain boundary property/ CSL behaviour generally carried out on pure materials. Investigations on alloys are usually in response to specific performance-related phenomena of industrial importance, such as boron-associated ductilisation of Ni_3Al[98] or growth of Goss grains in Fe-3%Si[99] (see Section 6.1.3).

4.3 SAMPLE POPULATION SIZES

In a grain boundary geometry investigation boundaries are selected manually for inclusion in a sample population, except for automated EBSD experiments. Typically a 'control' set of boundaries is sampled, plus other datasets reflecting different conditions concerning the property of interest, e.g. a variety of heat treatments. The property-affected boundaries may be a subset of the total dataset, which can then be extracted by post-processing. For statistical reliability, clearly larger sample populations are preferred.

Figure 4.2 shows the frequency distribution of sample sizes, indicating that many investigations encompass 200 boundaries or less. This relatively small number is a consequence of the fact that experimental techniques such as TEM or X-ray militate against large numbers of measurements. The larger datasets have been obtained using SAC or EBSD, and it is interesting to note that so far only three of these use fully automated EBSD. In other words, it is possible, but inconvenient, to collect quite large datasets by manual or semi-manual methods. Sample population sizes are presented as a function of CSLs in Figure 4.3, to show that there is no obvious connection between these two parameters.

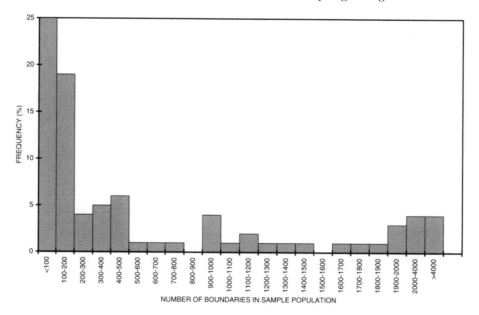

Figure 4.2   Distribution of CSL investigations on polycrystals according to sample population size.

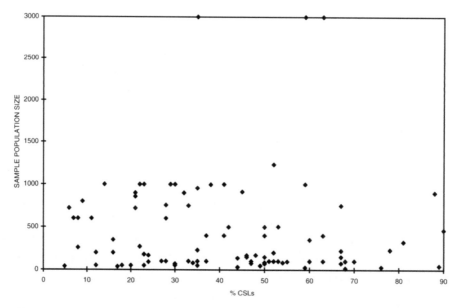

Figure 4.3   Proportions of CSLs v. sample population size, where this was reported.

4.4 MAXIMUM VALUE OF Σ

The maximum value of Σ to choose as a cut-off when calculating proportions of CSLs has been a long-standing enigma. Essentially, there is no physically correct answer to this question, since it implies a simple (and monotonic) connection between Σ-value and properties which is not the case in reality.[4] We are left, then with empirical choices. Although there is some evidence for special properties associated with 'higher' Σ-values, e.g. 31–61,[9] it is generally assumed that on the whole most special property boundaries will have lower Σ-values than this and moreover most CSL distributions show far fewer boundaries with 'higher' Σ-values than low-Σs. Figure 4.4 , which indicates the choices that have been made for maximum Σ, reflects this trend. There are outstanding maxima on Figure 4.4 at Σ29 and Σ49, that is, most workers have selected one of these values as a cut-off. The higher value is more relevant where orientation relationships are being discussed rather than the property of a boundary. This is because there is usually an attempt to correlate properties with low Σ-values, whereas orientation relationships relate more to geometry than properties (see Section 6.1.2). Strictly, the different values of maximum Σ selected invalidate comparisons of total CSLs between different investigations. However,

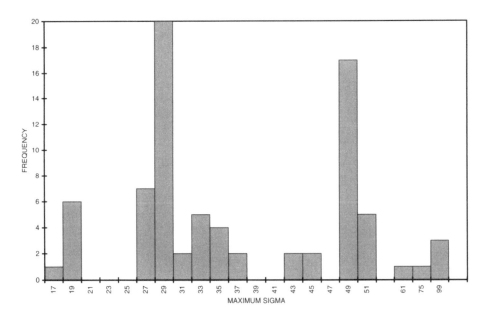

Figure 4.4   Frequency distribution of maximum Σ-values used in CSL investigations on polycrystals.

since in practice there are usually few CSLs which have higher Σ-values, qualitative comparisons are still effective.

4.5 GRAIN SIZE

It has been proposed that an inverse relationship exists between the proportion of CSLs and grain size for a range of materials.[100] The data systematised in the present study afford an opportunity to test this hypothesis, for investigations where grain size information is quoted in the original publications. Figure 4.5 shows the distribution of average CSL proportions and grain sizes. For plotting convenience grain sizes shown as 1000 μm also include those greater than this value. It is apparent from this analysis that there is no universal rule linking CSLs to grain size – and indeed there is no physical reason why there should be. An interesting point from Figure 4.5 is that more than half the investigations were conducted on materials with grain sizes less than 100 μm. This reflects partially the experimental techniques used, but more significantly that many materials of commercial importance have grain sizes in this range.

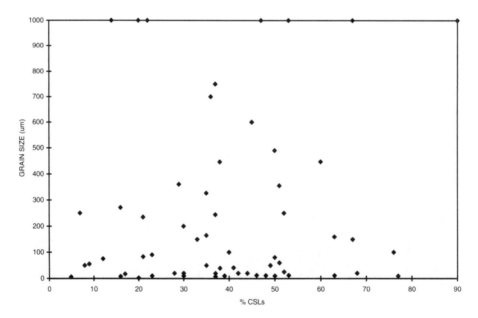

Figure 4.5   Proportions of CSLs v. grain size, where this was reported. The largest grain size shown, 1 mm, subsumes larger grain sizes.

4.6 CSL STATISTICS

The frequency distribution of CSLs for all the investigations is shown in Figure 4.6. As discussed in Section 4.4, although there are variations in the upper limit for Σ in the data used to compile this figure, there will not have a large effect on the statistics because in most distributions there are more CSLs with low Σ (i.e. ≤29, see Figure 4.4) than higher Σ (i.e. >29). The trend in Figure 4.6 approximates to a skewed normal distribution with the mode value for CSLs in the 40–50% group. The mean proportion of CSLs is 41%.

The global CSL statistics have been categorised further according to material, as shown on Figure 4.7. The materials are arranged in ascending order of CSL proportion, which highlights certain tendencies in the data, e.g. that there are fewer investigations on bcc materials than on fcc materials. The high average number of CSLs in copper, 66%, compared to the other materials, requires some elaboration. There are only five investigations comprising this category, and of those five, two report 80% and 89% CSLs respectively. These high proportions introduce a bias to such a small sample population on copper. The investigations refer to a study of grain boundaries for nuclear waste storage, i.e. an example of grain boundary engineering,[101] and a one-dimensional polycrystal.[75]

The most striking aspect of the statistics in Figures 4.6 and 4.7 is the

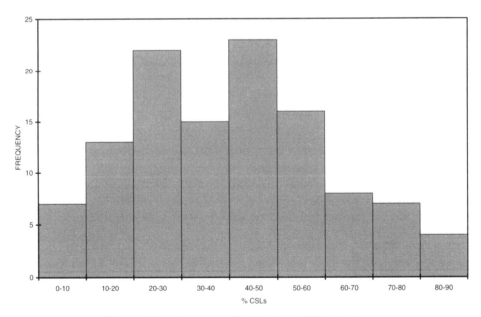

Figure 4.6   Frequency distribution of CSL fractions.

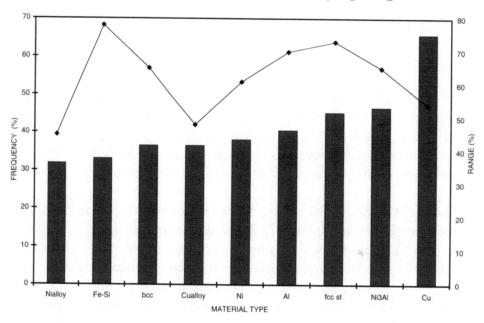

Figure 4.7   Average CSL statistics per material group. Both the mean % CSLs (histogram) and the % range (diamond symbols) is included for each group.

comparison between the measured proportions and the proportion of CSLs expected for a random distribution of grain boundaries. Reiterating from Section 2.2.3, only 11% of grain boundaries are expected to be CSLs having $\Sigma < 27$ (including $\Sigma 1$s) in a random distribution of grain pairs. If we compare this baseline figure with that illustrated in Figure 4.6, it is clear that almost all investigations report CSL fractions many times greater than random. Frequently the same material can exhibit a large range of CSL proportions, depending on the processing stages and other factors such as impurity content. CSL proportions in the same material type vary from about 10% to 70% (Figure 4.7).

The two factors which have been observed to exert the most influence on the number of CSLs in a sample are

• The presence of a strong texture, resulting in $\Sigma 1$ boundaries;
• The amount of twinning, resulting in $\Sigma 3$ and $\Sigma 3''$ boundaries.

A strong texture, especially a single-component texture such as the cube or Goss, introduces $\Sigma 1$s into the system because there is an increased probability that grains having almost the same orientation will reside as neighbours. The presence of many subboundaries in the system will also increase the $\Sigma 1$ fraction. Low or medium stacking fault energy materials, notably many austenitic steels or copper, are highly susceptible to twinning which increases the $\Sigma 3''$ fraction.

The average proportion of CSLs does not, of course, indicate their distribution. Two materials which have the same average proportion of CSLs may yet have entirely distinct distribution profiles. This is illustrated by a recent experiment which uses OIM to measure the proportions of CSLs in two different materials: Inconel 600, a nickel-based alloy which has a weak texture and aluminium thin film which has a strong fibre texture.[85] Although the overall percentage of CSLs is similar in each (>50%), the division of CSL types is quite different since the twinning mechanism governs the distribution in the nickel alloy whereas the fibre texture in the aluminium has resulted in a large population of $\Sigma 1$s. The distribution of these CSLs is shown in Figure 4.8, including computer simulated distributions generated without consideration of connectivity.

Figure 4.9 gives more insight into the origins of average CSLs for each material by showing the average proportions of $\Sigma 3$s for each material group (Figure 4.9a) and the ratio of $\Sigma 1$s/$\Sigma 3$s (Figure 4.9b). It should be noted that not all investigations reported $\Sigma 3$ and $\Sigma 1$ fractions separately, and compilation of the data in Figure 4.9b was therefore restricted to cases where these figures were available. The amount of $\Sigma 3$s in the fcc categories correlates to a first-order approximation with the stacking fault energy. This is illustrated with reference to Table 4.1 which lists values of stacking fault energy and average proportions of twins collated in the present analysis. Turning to the $\Sigma 1$ category, proportions of $\Sigma 1$s are most significant for bcc materials, and aluminium. This can be explained by the fact that bcc materials and aluminium often exhibit strong textures, and furthermore annealing twinning is not expected.

To summarise this section, experimentally observed proportions of CSLs are frequently far in excess of those expected on the basis of a random distribution. $\Sigma 1$ and $\Sigma 3$ are the most prevalent types, caused by strong textures and twinning respectively. Since repeated twinning tends to

Table 4.1   Approximate stacking fault energies (SFE) and average twin proportions for fcc materials in this study

Material	SFE (mJm^{-2})	%$\Sigma 3$
Al	170	5
Ni	130	28
Cu	80	45

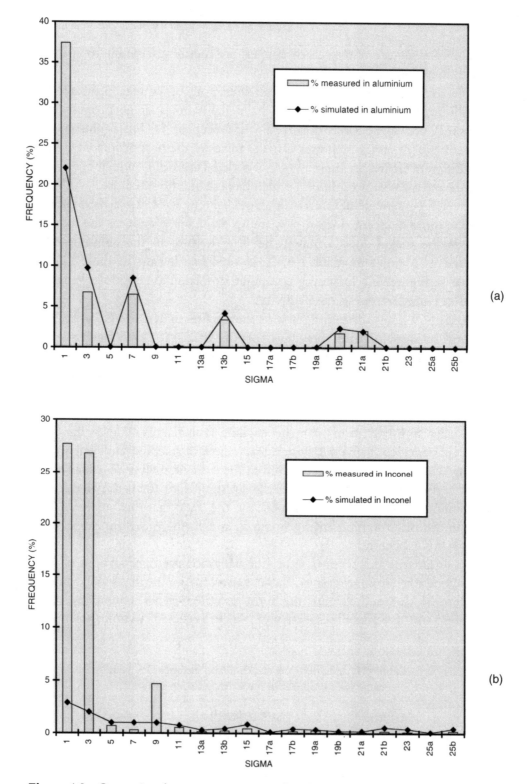

Figure 4.8 Comparison between computer predicted and measured proportions of CSLs in (a) alumimium and (b) Inconel.[85]

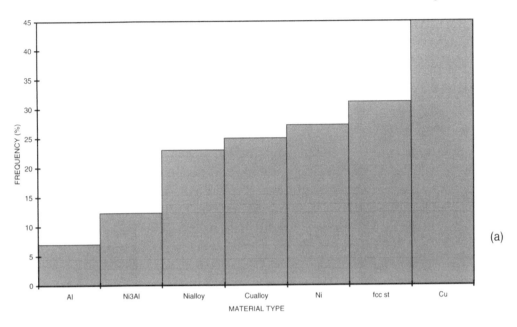

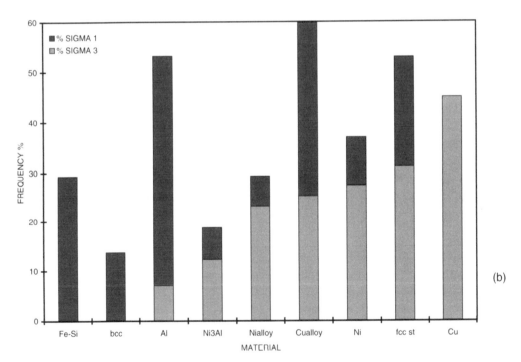

Figure 4.9   (a) Average proportions of $\Sigma 3$ boundaries per fcc material. (b) Average proportions of $\Sigma 3$ boundaries and $\Sigma 1$ boundaries per material where both these statistics were quoted in an investigation.

randomise the texture, high proportions of Σ1s and high proportions of Σ3s tend to be mutually exclusive.

4.7 CONTINUOUS FUNCTIONS

Use of continuous functions to display grain boundary statistics provides a connection between traditional texture research and grain boundary research. The misorientation distribution function (MODF) is the most common continuous function used to represent grain boundary data. The 'uncorrelated' MODF describes the probability that a grain boundary separates grains of a certain misorientation. This can be computed from X-ray generated macrotexture data, i.e. an orientation distribution function, ODF, or from individual grain orientation measurements.

The MODF becomes a more powerful analytical tool when it is constructed from the real spatially specific orientations that exist in the microstructure. This is called the 'physical' MODF.[81] A further function is defined, the 'orientation correlation function', obtained when the physical MODF is divided by the uncorrelated MODF. In other words the orientation correlation function is a 'filtered' distribution, with effects due to random chance occurrences of neighbouring grains – the uncorrelated MODF – removed. Other terms may be used for the MODF heirarchy, and these are documented elsewhere.[17,82,83,102] MODFs can be displayed in Euler space, (which is popular where the measurements have been generated from X-ray data), stereogram-based space[17] or Rodrigues-Frank (RF) space.[103,104] Peaks arising in MODFs can be identifed in angle/axis notation and hence CSL data can be extracted as shown in Figure 4.10 where some peaks are labelled.[105] A strength of the MODF approach is that all misorientation peaks are apparent, whether or not they are CSLs. For example, in Figure 4.10 there is a strong peak at $\alpha = 70°$, $\beta = 75°$, $\gamma = 20°$ which is not a CSL.

Misorientations are represented by only one of their 24 symmetry-related variants, as described in Section 2.2.3. Hence the space in which an MODF is displayed can be diminished to 1/24 of its original volume, which is particularly apposite for CSL research since most CSLs lie on the surface, including some edges and corners, of the reduced domain. This is called the 'asymmetric domain'[84,106] and the 'subvolume of the fundamental zone'[103] in Euler space and RF space respectively. The form of these two subspaces is shown in Figure 4.11, and it can be seen that RF space is geometrically the more simple of the two spaces since it has planar faces. Another key difference between the Euler and Rodrigues parameters is that the sub-

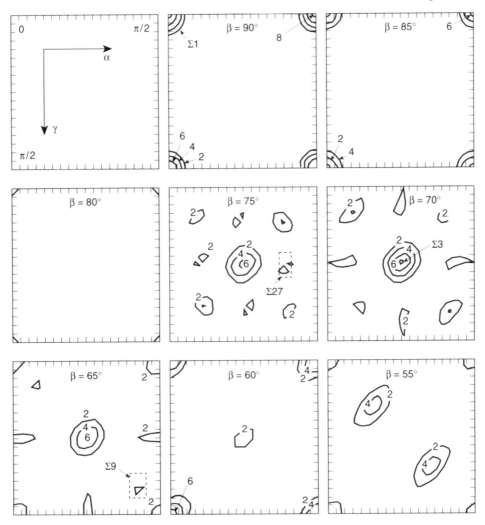

Figure 4.10 Intercrystalline structure distribution function (ISDF) in Euler space for a sample population of boundaries in an fcc steel. The contours are expressed as times random. CSL peaks are labelled.[105]

volume of RF space contains the misorientation which represents the lowest angle solution of the 24 variants, whereas this may not the case for the asymmetric domain of Euler space.

The concept of 'multiplicity' arises in connection with positions of misorientations on the surface of the domain/subvolume. The multiplicity defines the number of physically indistinguishable positions in the whole misorientation space.[107] A multiplicity $m = 1$ signifies that the misorientation has no particular symmetry, therefore there are 24 distinct ways to express

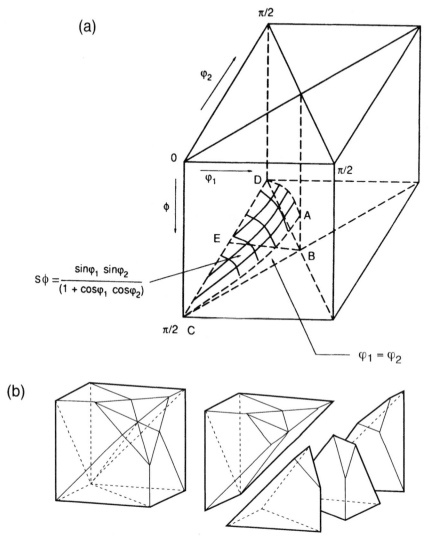

Figure 4.11   (a) Asymmetric domain in Euler space[107] (b) Octant of Rodrigues-Frank space partially exploded to show the six fundamental zones which it contains. (Courtesy A. Day.)

the angle/axis pair. For these cases the misorientation lies within the domain. Higher multiplicities mean that the misorientation lies on the surface ($m = 2$) or the edge of the domain ($m > 2$). All CSLs having $\Sigma \leq 49$ lie on the surface or edge of the domain, except for $\Sigma 39b$ which has a multiplicity of 1. $\Sigma 3$ is a special case in that it lies on an apex of the domain, i.e. where edges meet. These points have been clarified here because it has been postulated that grain boundaries which lie on the surface of the domain (which includes

CSLs) are a special class of interfaces which have shown preferential interfacial damage during creep experiments, except for $\Sigma = 3$, which lies at the apex of the domain.[108]

Continuous functions have been extended to more sophisticated analyses of grain boundary geometry. These include the intercrystalline structure distribution function, ISDF,[83] which describes the boundary in terms of five parameters and the interface damage function, IDF,[84] which considers the fraction of damaged (e.g. cavitated) boundaries in a sample population. These functions can be extended to grains which are not nearest neighbours and can be analysed using CSLs as a reference frame, similarly to the methodology for MODFs.

4.8 COMPUTER MODELLING

It is a relatively simple procedure to explore CSL distributions via computer simulation of grain orientations followed by calculation of relative misorientations. This methodology has been adopted by several workers in the past few years, particularly to examine the effect of texture on CSL distributions in polycrystals and to gain more insight into the factors which control CSL generation.

The simplest case is a simulation of non-textured orientations, which is obtained using a random number generator. Misorientations for typically many thousands of simulated grains pairs are collated and the CSL fractions calculated. Some investigations have included more complex grouping of grains.[86,87,88] All the work referenced in the present section has used the Brandon criterion to establish the CSL fractions. A second approach for assessment of CSL fractions is to use a probability algorithm to calculate expected CSL proportions, and these have been shown to agree with computer generated results (Figure 4.8).[85]

The combined computer generation/analytical approach was first used in 1975 to give expected proportions of CSLs, as mentioned in Section 2.2.3.[39] Since then, other workers have conducted similar experiments to deduce the proportion of CSLs associated with random texture. These results are shown on Figure 4.12, where it is clear that both the trends and absolute numbers are in good agreement. The key point is that the proportion of CSLs is small if the texture is random – not much more than 2% and 1% for $\Sigma 1$ and $\Sigma 3$ respectively.

A theme which recurs frequently in simulation experiments is that *texture*

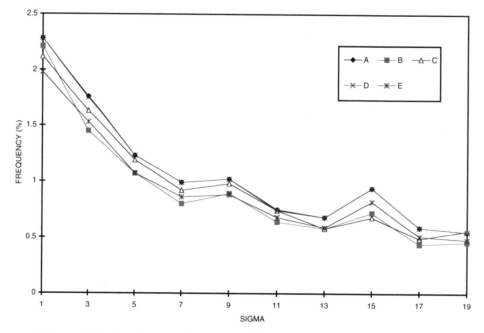

Figure 4.12   Distribution of CSLs up to $\Sigma 19$ for random texture obtained from computer modelling by several different workers: A,[85] B,[86] C,[87] D,[116] E.[39]

increases the CSL fraction. In particular, strong fibre textures, 100, 110 and 111, have been imposed on simulated orientations which augment the fraction of CSLs having misorientation axes parallel to the fibre axes.[85,86,90,109,110,111] $\Sigma 1$s are always increased. Whereas these broad conclusions could have been predicted intuitively – e.g. if many grains have a 100 texture there will be many grain pairs having a common 100 misorientation axis – simulations have the benefit of providing quantitative data. Simulations on common two-dimensional textures (e.g. brass, Goss, copper, etc) have also been carried out.[112] These investigations have shown that the proportion of CSLs generated has essentially the same characteristics for textures modelled in tricrystals (to simulate triple line geometry) and those modelled in two-dimensional, connected polycrystals. A further subtlety found was that texture modifies the misorientation angle more than it does the axis.

Several simulations have been designed to make comparisons with experimental measurements in real materials. For example, the trends of the CSL distribution in aluminium film having a strong 111 fibre texture are reproduced accurately by simulation as shown in Figure 4.8. Quantification of the increase in those CSLs having misorientation axes corresponding to the fibre axis has been studied several times by simulations, and show agreement with the microstructure.

Another simulation approach is to forecast those CSLs which are preferred as orientation relationships between deformed/recrystallised or primary/ secondary recrystallised components of microstructure. For example, an unexpectedly high occurrence of Σ35b CSLs across recrystallisation inter-faces in an austenitic steel was explained by the relationship between variants of the principal texture components {1 1 0}<1 1 2> (deformed) and {3 1 1}<1 1 2> (recrystallised), and furthermore between the deformation component and twins of the recrystallised component.[88] The Σ35b peak is shown in Figure 4.13. Aspects of orientation relationships are discussed further in Section 6.1.2.

Although for the two cases discussed above (i.e. fibre textures and orientation relationships) there is fairly good agreement between simulation and experiment, where there is a high level of twinning in a material a large discrepancy can exist between simulation and experiment, unless the twinning is specifically incorporated in the simulation. This is because twinning tends to randomise texture while increasing concurrently the number of $\Sigma3^{n}$ and possibly $\Sigma1$ boundaries in the polycrystal. Some

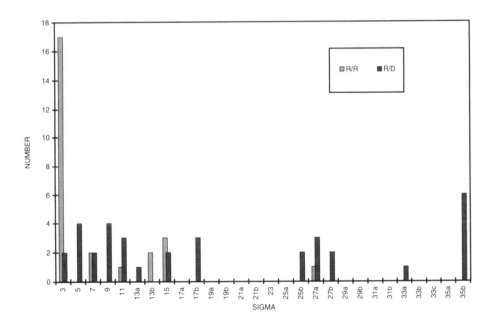

Figure 4.13 Distribution of CSLs having $3 \leq \Sigma \leq 35$ for boundaries between recrystallised and deformed grains (R/D) and boundaries between newly recrystallised grains (R/R) in an austenitic steel.[88]

simulations have incorporated twinning and its effect on the CSL distribution.[113] More recent simulation work takes account not only of the CSL designation of boundaries but also of the CSL characteristics at grain junctions.[90,112] This is discussed in Section 5.2.2.

To summarise this section, computer simulations of misorientations underpin grain boundary experiments on real materials by providing a theoretical baseline of expected CSL proportions e.g. for random orientations. However, there are limitations to this approach since there is not always agreement between simulation and experiment, particularly where short-range grain correlations, e.g. twinning or clustering, exist. There has tended to be a chronological evolution of modelling sophistication and usefulness to real materials, as progress has been made from simulations of random orientations to fibre texture, full textures, orientation relationships and most recently grain junctions.

4.9 THE COINCIDENCE SITE LATTICE AND SPECIFIC BOUNDARY PHENOMENA

The distribution analysis of all the properties or features recorded in the literature set on CSLs in polycrystals is shown in Figure 4.14. In this context 'grain boundary features' have been divided into orientation relationships/ phase transformations, grain connectivity, grain boundary statistics, microstructural evolution, texture and twinning. Some investigations cover more than one property or feature, and for these cases the major property has been included in Figure 4.14. This digest highlights two points: some properties are experimentally more feasible to measure than others, and similarly some properties are of especial commercial/research interest compared to others.

It is instructive to consider the range of mechanical, physical and chemical properties attributed to grain boundaries compared to the properties/features which have actually been studied. Most of the properties mentioned in Section 2.3, namely cavitation, corrosion, diffusivity, embrittlement, energy, fracture, migration, mobility, precipitation, resistivity, segregation and sliding, have been studied in polycrystals, although necessarily not in the controlled manner possible for bicrystals where exact CSLs can be prepared. Bicrystal data provide a framework and a guide to special properties for well-defined cases; however they do not extrapolate to polycrystals in a straightforward manner, as discussed in Section 3.1. Experiments to explore the attributes of CSLs have therefore needed to be adapted for application to the general polycrystal case.

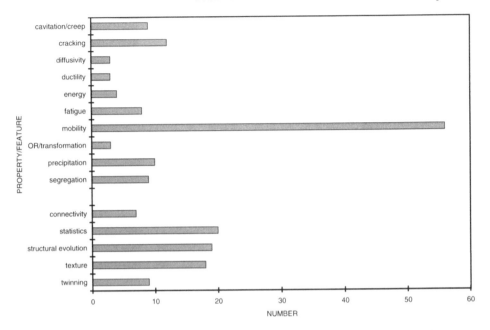

Figure 4.14 Distribution of specific CSL investigation topics. Where more than one property or feature is analysed in a reported investigation, the principal one only is included in the distribution.

A typical experimental adaptation for polycrystals is that comparisons are made globally between entire sample populations having different features (such as solute content) rather than individual boundaries within a specimen. It is also clear from Figure 4.14 that the CSL investigations which have been performed embody areas of structure/property relationships which are prominent in research and technology – for example, recrystallisation, secondary recrystallisation, evolution of microstructure, relationships between boundaries and intergranular degradation associated with fatigue, creep and corrosion.

In the remainder of this section, the rationale employed to study grain boundary properties/features in terms of CSLs will be analysed in terms of common trends in the reported data.

4.9.1 Mobility

Nearly one-third of the reported investigations are concerned with which CSLs are the most mobile – or more accurately which CSLs migrate first/fastest under specific experimental conditions. Because of its predominance in Figure 4.14, the mobility group has been extracted from the other properties for separate consideration. The group subdivides into recrystallisation or grain growth. The most typical experiment in the former subgroup is performed on part-recrystallised TEM specimens of pure metals or dilute binary alloys.[114] The CSL scheme is used principally as a classification tool to sort the results. Frequently a large tolerance is permitted on the CSL classification, e.g. up to 10°.[115] The motivation for this type of experiment is to elucidate factors controlling recrystallisation behaviour (see Section 6.1.2).

The majority of the grain growth group involve secondary recrysallisation, particularly in Fe-3%Si transformer steel: 21 of the 56 reported mobility investigations were performed on this material. This large fraction reflects the high level of technological importance since formation of a strong Goss texture $\{1\,1\,0\}<0\,0\,1>$ is crucial to development of optimum magnetic properties. The Goss texture develops by means of intense secondary recrystallisation, which in turn is the result of very selective grain boundary migration. In grain growth experiments the CSL scheme is used to analyse, and in some cases predict, the boundary migration characteristics (Section 6.1.3).

Grain boundary migration experiments, both across recrystallisation interfaces or after recrystallisation are among those which are easy to perform. The most frequently adopted methodology involves collecting grain boundary statistics before and after grain boundary migration and comparing the proportion of CSLs in the two data sets. Additionally sample populations may be divided into subsets of, for example, recrystallisation interfaces/ boundaries between newly recrystallised grains, or boundaries between anomalous and small grains and other control data. A drawback to using this type of experiment to gain information about grain boundary migration is that the driving force will be affected inevitably by the free surface energy.

4.9.2 Other properties

Besides mobility, other grain boundary properties which have been investigated are cavitation/creep, cracking, diffusivity, ductility, energy, fatigue, precipitation and segregation. The two most common ways for the property measurement to be approached are for the CSLs in a specimen displaying the particular property to be compared with a control specimen, or for the property to be observed/measured at individual boundaries. Examples of the control specimen approach are the comparison between a ductile and a non-ductile specimen, e.g. Ni_3Al,[98] or comparisons between specimens containing segregants e.g. sulpur[116] phosphorus[117] boron[118] and pure specimens. Examples of the observation/measurement approach include precipitation density quantification at individual boundaries in steel[119] or fatigue-induced cavitation along individual boundaries.[120] All of these investigations attempt to correlate the occurrence of CSLs with particular properties. Most of the reported investigations indicate that there is a relationship between CSL fraction and property, although the correlation does not tend to approach 100%. Evidence for this statement is shown in Table 4.2, which summarises a range of experiments encompassing various types of intergranular degradation (corrosion, cavitation, fracture) and segregation or precipitation, labelled C and S respectively. Only those investigations which report unambiguous statistics are included. Table 4.2 shows that only $\Sigma1$ and $\Sigma3$ coherent twins are universally reported to have special properties; in fact in some investigations these are the only CSLs which correlate with special properties.[121] More generally, decreased intergranular attack or segregation effects correlate with an increased proportion of CSLs.

It is important to realise that the CSL/property correlation, particularly when related to intergranular attack, depends on the agressiveness of the environment and furthermore depends partially on factors external to the intrinsic structure of the boundaries themselves. An example is recent work on copper crept at $0.6T_m$ for 0.6 of the fracture life.[108] Here the only boundaries resistant to creep damage were coherent twins and $\Sigma1$s where the misorientation angle was <10°. All other CSLs behaved as general boundaries under these quite severe conditions; boundaries with multiplicities >1, which includes CSLs, were damaged more than boundaries with multiplicities = 1 (see Section 4.7). Most of the damage occurred on boundaries which were oriented nearly normal to the principal stress axis, indicating that this factor overrode other considerations.

The investigations encompassing CSLs/grain boundary energy have been carried out in polycrystals under conditions where the boundaries possess some rotational freedom, e.g. in a wire or across a corner of a specimen.[75,122] Apart from these special conditions, it is not straightforward to obtain grain boundary energy directly in a polycrystal. One other feasible method is from dihedral angle measurements at the intersections of coherent twin boundaries with high-angle boundaries.[123,124] Similarly, diffusivity is not trivial to measure in polycrystals. The reported experiments concerning the relationship between diffusivity and the CSL are based on the 'disappearance', i.e. absorption, of dislocations into grain boundaries, and are performed using a hot stage in the TEM.[59] The simplified expressions which govern the absorption of EGBDs into a boundary and allow the diffusivity to be calculated from the measurable parameters are:[5]

$$t_D = AT_D/D_b\delta \qquad (4.1)$$

$$Q_b = RT_D \log Kt_D/T_D \qquad (4.2)$$

Table 4.2   Summary of investigations relating to grain boundary degradation (C) and segregation or precipitation (S).

Property	Material	Description of results	Reference (see appendix)
C	fcc st	Reduced cracking at Σs 5, 17, 13, 1. Columnar grains with a strong 100 texture.	Sato et al (1986)
C	Cu	Cavitation did not correlate with CSLs	Adams et al (1990)
C	Alloy 600	69% CSLs Σ >29 cracked. 56% CSLs Σ <29 uncracked	Aust et al (1993)
C	Alloy 600	Increasing CSLs from 35% to 67% reduces corrosion rate	Lin et al (1995)
C	Cu	Cavitated boundaries: 10% CSLs, 31% non-CSL; Non-cavitated boundaries: 38% CSLs, 21% non-CSL	Butron-Guillen et al (1990)
C	Alloy 600	Treatments giving 35–45% CSLs have less cracks than those giving 12–20% CSLs	Crawford & Was (1992)
C	fcc st	8% CSLs cavitated – Σ1 and Σ3 all uncavitated; 15% non-CSL boundaries uncavitated	Don & Majumdar (1986)
C	Al-Li	92% cavitated boundaries are non-CSLs	Randle (1995)
C	Ni$_3$Al	Specimens with 50% CSLs ductile (including 26% Σ1s) Specimens with 28% CSLs brittle	Farkas et al (1988)
C	Cu	Cavitation on all boundaries except coherent Σ3s and Σ1s <10°	Field & Adams (1992)
C	Cu/Al$_2$O$_3$	All CSLs unfractured plus a few non-CSLs One-dimensional grain structure	Jaeger & Gleiter (1978)
C	Cu	Cavitation on all boundaries except coherent Σ3s. Preferential damage on boundaries on surface of asymmetric domain	Field (1995)
C	Ni	Non-CSL boundaries cavitate first. Eventually all boundaries cavitate except Σs 3, 1, 5	Lim (1987)
C	Ni$_3$Al	Only coherent Σ3s and Σ1 uncracked	Lin & Pope (1993)
C	Ni	CSLs with Σ <27 mainly resistant to corrosion, depending on the severity of attack	Palumbo & Aust (1988)
C	Ni$_3$Al	Only Σs 3, 1 fracture resistant	Hanada et al (1986)

Table 4.2 – *Continued*

Property	Material	Description of results	Reference (see appendix)
S	bcc st	Low P segregation at $\Sigma3$s and $\Sigma1$ only	Ogura et al (1987)
S	Ni$_3$Al	Segregation of B has no effect on the CSL distribution	Mackenzie et al (1988)
S	fcc st	No precipitation on coherent $\Sigma3$. Low precipitation on Σs 1, 9, 27. Precipitation on other boundaries	Liu et al (1995)
S	Ni-S	Increasing S content has no effect on the CSL statistics	Palumbo & Aust (1990)
S	fcc st	No Cr segregation to Σs 3, 11, 13a, 13b. Cr segregation to all other CSLs and non-CSL boundaries.	Laws & Goodhew (1991)
S	fcc st	85% CSLs have no precipitates 23% non-CSLs have no precipitates	Randle and Ralph (1987b)
S	fcc st	No corrosion at 57% $\Sigma3$s, 27% non-CSLs All coherent $\Sigma3$s and $\Sigma1$ are corrosion resistant	Ortner & Randle (1989)
S	Fe-P	Increase in CSLs for specimens with reduced phosphorus or increased recrystallisation temperature	El M'Rabat & Priester (1988)
S	Ni-S	No segregation of S to CSLs except $\Sigma9$	Bouchet & Priester (1987)

where D_b is the grain boundary diffusion coefficient, δ is the grain boundary width, Q_b is the activation energy for grain boundary diffusion, T_D and t_D are the temperature and time for which dislocation contrast disappears in the TEM, R is the gas constant and A, K are constants.

4.9.3 Grain boundary features

Turning now to the other characteristics of grain boundaries which have been related to the CSL model, these are: connectivity, orientation relation-ships/phase transformations, statistics, structural evolution, texture and twinning as shown on Figure 4.14. The connectivity group takes a more physically realistic strategy to the network aspects of grain boundaries in polycrystals. This is done either by a correlation function approach[72] or by analysis of grain boundary conjunctions at triple lines.[90,91,92] The CSL scheme has also been used to describe the orientation relationships occurring during phase transformations, most commonly the austenite-ferrite transformation in steels.[125]

Inclusion of an investigation in the 'statistics' category denotes that the primary motivation for the study was to obtain the distribution of CSLs (usually to compare with that for a random distribution) between sample populations or with computer-generated data. 'Structural evolution' is a category intended to subsume those investigations which deal with how the CSL fraction is linked to changes in the microstructure, usually as a function of thermomechanical processing. For example, changing heat treatment schedules in nickel has been shown to alter radically the CSL distributions.[126]

Texture (preferred orientation) is an important area of study in its own right, and where data have been collected on a grain-by-grain basis (i.e. microtexture), can be integrated with CSL analyses. In some experiments this is achieved by calculating the misorientation between peaks in the orientation distribution[127] or by calculating the 'physical' or 'statistical' MODF (see Section 4.7). Another approach is to calculate or model the CSLs generated from specific orientations and use this information to analyse experimental data. This approach has been applied to, for example, recrystallisation in IF steels.[128]

Twinning has major significance in CSL-based analyses of polycrystals. As discussed in Section 2.3.2, an annealing twin is a $\Sigma 3$ boundary and, for the coherent type (i.e. where the boundary plane is 1 1 1 indexed in both grains) the energy is extremely low, typically about $20 \, \text{mJm}^{-2}$ compared to about $900 \, \text{mJm}^{-2}$ for a general high angle boundary. The value for the coherent twin is at least an order of magnitude lower than that for other low-energy CSLs. Twinning is therefore a means of introducing low-energy boundaries – whose other properties such as diffusivity are also markedly special – into a system. By so doing the material can be 'grain boundary engineered' to give improved properties. The exploitation of twinning has been one of the cornerstone successes for application of the CSL model and grain boundary engineering,[12] and for this reason it is discussed in more detail in Section 6.1.4.

5. *Extensions of the Coincidence Site Lattice*

5.1 GRAIN BOUNDARY PLANES

5.1.1 Introduction

This section addresses the extension of CSL grain boundary analysis to inclusion of the grain boundary plane. The formalism for the classification of grain boundary planes within the CSL hierarchy was discussed in detail in Section 2.2. To reiterate briefly, the most salient point was that the misorientation provides only three of the five independent variables describing grain boundary geometry; the other two refer to the orientation of its plane. The boundary surface, where it is actually planar, is described crystallographically by its indices referred to the crystal axes of both interfacing grains.

In a polycrystal grain boundary surfaces can in principle take up a wide range of positions which are governed broadly by conservation of the grain shape and equilibration of surface (grain boundary) energies. On average, a grain has fourteen surfaces and each one of these represents a crystallographic plane. Using microscopical or other techniques, the inclination of those boundaries which intersect with the plane of polish in a specimen can be measured. If this information is coupled to diffraction information from grains, the crystallography of planar boundaries can be obtained (see Section 5.1.4). The motivation for such an exercise is to gain more information about the grain boundary structure than is revealed from knowledge of the misorientation alone.

5.1.2 Theoretical considerations

The total boundary crystallography is categorised according to the schedule in Figure 5.1. The first categorisation is whether or not the planes are related to a CSL. Up to two STGBs are possible geometrically per CSL and an unlimited number of ATGBs per CSL, which for practical purposes is usually limited to quite low Miller indices. The number of TWGBs may be up to 24 depending on the geometry of the particular CSL. The type of tilt/twist boundary has implications for the free volume and hence properties of the boundary. STGBs have low free volume compared to a general

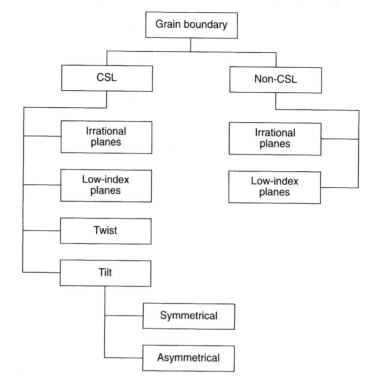

Figure 5.1   Flow chart showing the hierarchy of possible grain boundary types in the interface-plane scheme.

boundary, although it has recently been realised that STGBs do not necessarily have the *lowest* free volume/energy in a CSL system.[49] ATGBs and TWGBs also have lower than average free volume, some values of which have been estimated by computer simulation methods.[50] On the whole tilt boundaries have lower energies than twist types, and ATGBs in some CSL systems have lower energies than STGBs in other systems. These points are discussed further in the following section.

Where the boundary planes are not related to the CSL, i.e. the boundary is neither a twist nor a tilt type, it may still have rational, low-index planes with respect to one or both of the interfacing grains as indicated on Figure 5.1. For example, microfacets close to 1 1 1 and 1 1 0 with respect to both interfacing grains respectively have been observed in gold using HREM. These planes do not conform to an ATGB in any CSL system. However, this incommensurate boundary is 1.14° from a Σ41 tilt having 24,24,23/001 boundary planes. (This notation refers to the Miller indices of the grain boundary in both interfacing grains, i.e. 24,24,23 in the first grain and 0 0 1 in the second for this case). The high resolution micrograph in Figure 5.2

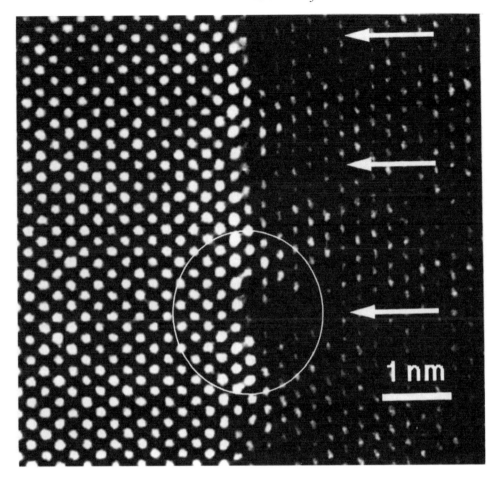

Figure 5.2 HREM image of a 111/001 grain boundary in gold, with quasiperiodic structural units indicated by arrows. One such unit is indicated by a circle.[129] (Courtesy K.L. Merkle.)

shows evidence that several atomically well-matched regions along the boundary are separated by regions of misfit localisation.[129] The implication is that the 111/110 configuration is more favourable energetically than the nearby configuration which preserves the $\Sigma 41$ CSL.

Experiments have shown that the grain boundary may be characterised by irrational indices with respect to both grains, whether or not it has been designated a CSL on the basis of misorientation.[130] For these cases it is unlikely that the periodicity benefits of a CSL are realised in terms of low free volume, and the boundary is unlikely to show special properties.

The inference from the above description of boundary plane categorisation is that even if a boundary has a low Σ-value, this does not guarantee that it will exhibit special properties because the plane may not be in a favourable location. Several workers have emphasised the critical importance of knowing the indices of the grain boundary plane,[5,21,106,131] yet the present work has shown that only about 10% of reported investigations, compared to the total population of CSL experiments, have included the plane as a measured parameter. In this Chapter the information gleaned from those experiments which have included boundary plane data in addition to routine CSL measurements will be critically appraised.

5.1.3 Grain boundary energy and boundary planes

Free volume can be related to the boundary plane by introducing the parameter d/a, where d is the interplanar spacing of the boundary planes for a STGB normalised by the lattice parameter a.[132] On this basis STGBs which correspond to energy cusps also correspond to high d/a values which in turn are matched by the closest-packed planes in the CSL. Figure 5.3 demonstrates this relationship by showing the energies and STGB planes of some CSLs.[48] The 1 1 0 and 1 0 0 boundary planes correspond to $\Sigma 1$. The d/a approach is equivalent to the identification of favoured boundaries having only one type of structural unit in the boundary.[35]

It has been suggested that the d/a criterion for a low volume/energy boundary can be adapted to ATGBs by averaging the interplanar spacing of

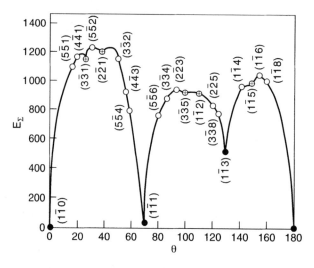

Figure 5.3   Grain boundary energies (mJm^{-2}), as determined by a simulation procedure, for <1 1 0> STGBs as a function of misorientation angle. 'Special' orientations, giving rise to primary cusps, are denoted by full symbols and crossed circles are used to indicate secondary cusps.[48]

planes on both sides of the boundary, d_1 and d_2, giving an effective value of d, d_{eff}:[48]

$$d_{eff} = (d_1 + d_2)/2 \qquad (5.1)$$

Although there has been some support for use of d_{eff} as a guide to 'special' boundaries, there are some drawbacks. For example, the grain boundary with the highest d_{eff} in an fcc structure is 1 1 1/1 0 0, which does not have a periodic (i.e. CSL) geometry. However, if the criterion for low energy is based on a high d_{eff} value, then this boundary would be expected to have been observed in rotating sphere experiments, which is not the case.[4] This difficulty may be rationalised from the evidence highlighted in the previous subsection, i.e. that periodic planes in the CSL may facet on an atomic level into regions of high density planes interspersed by steps.[129, 133]

Actual values for grain boundary energy can be obtained by calculation, simulation and measurement. Figure 5.4 shows boundary energies for copper and gold for selected tilt boundaries in the $\Sigma 3$, $\Sigma 9$ and $\Sigma 11$ systems.[50] Note that the absolute values may vary with material, although the trend remains almost the same. Comparisons between the values in Figure 5.4 illustrate the important point that the energies of certain ATGBs can be lower than an STGB for the same CSL. For example in the $\Sigma 9$ system, the

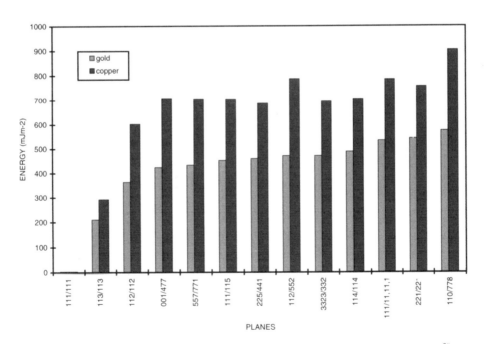

Figure 5.4 Calculated energies of <110> tilt boundaries for gold and copper.[50]

557/771 and 225/441 ATGBs both have lower energies than the 332/332 STGB.

The family of 110 tilt Σ3s is a particularly important set of boundaries because they have low energies and occur frequently via twinning in materials such as austenitic steels, copper and nickel. Both the 111/111 and the 112/112 STGB are composed of a single type of structural unit; ATGBs are composed of a mixture of these units. As 112/112 units are added to the 111/111 structure the boundary energy increases almost linearly with angular deviation from the 111/111 configuration as shown in Figure 5.5. The departure from linearity on this curve is explained by the incorporation of a thin rhombohedral phase at the boundary. Figure 5.5 shows that there are many Σ3 ATGBs which have lower energy than the 211/211 STGB.[134]

Qualitative energy rankings between different boundary planes are available from certain rotating sphere experiments. Evidence from copper and silver shows qualitatively a hierarchy of boundary energies. Table 5.1 shows those boundary planes which were observed to give very strong, strong or medium X-ray reflections.[135] The data in Table 5.1 further underpin the evidence that many ATGBs have relatively low energies. To summarise, in general tilt, twist and boundaries having low-index planes in both grains are likely to have low energies. In addition to energy, most other grain

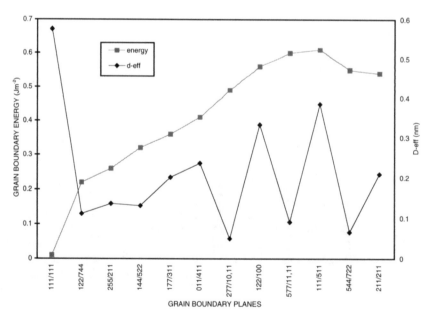

Figure 5.5   Relationship between calculated grain boundary energy[134] and d_{eff} for <110> tilt boundaries. Individual data points are joined as a guide to the eye.

Table 5.1 Intensity of X-ray reflections corresponding to boundary planes in a sphere-on-plate experiment with fcc metals (Cu or Ag)[67]

Planes	Σ-Value	Reflection
111/111	3	vs
411/110	3	vs
221/100	3	vs
511/111	3	s
110/110	3	s
766/100	11	s
511/111	9	m
19,1,1/111	11	m
13,13,5/111	11	m
37,37,23/111	33c	m
19,1,1/111	33c	m
8,7,7/110	9	m
13,8,5/110	11	m
40,40,17/110	33c	m

(vs = very strong; s = strong; m = medium)

boundary properties will be similarly affected by the boundary plane orientation.

5.1.4 Methods for obtaining grain boundary plane data

Since methods for obtaining boundary plane data are described in detail elsewhere[17,136] the topic will be discussed only briefly here. The input data for the determination are the inclination of the boundary plane and the orientation of each interfacing grain, all measured with respect to the *same* reference system. Figure 5.6 illustrates the non-crystallographic parameters. XYZ are the reference axes, α can be measured directly off a polished section and β can be obtained from:

- A two-dimensional specimen where boundary traces can be observed on both surfaces;[5]
- Calibrated serial sectioning;[136]
- Sectioning on two mutually perpendicular surfaces as illustrated in Figure 5.6;[122]
- TEM of highly inclined boundaries in the foil;[137]

Care must be taken to measure α and β in the correct sense with respect to XYZ. After the direction cosines of the boundary plane in the coordinate

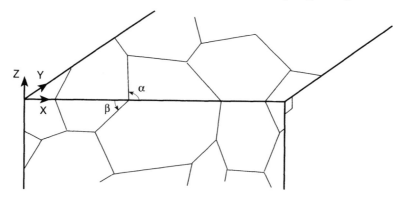

Figure 5.6   Schematic illustration of grain boundaries traversing a corner of a specimen and also the angles α and β which need to be measured in order to obtain the boundary inclination.[136]

systems of each interfacing grain have been calculated, these can be compared to a look-up table to check for tilt and twist configurations for the particular CSL.

5.1.5 Overview of experimental data

The principal results from experiments which determine explicitly the orientation of the boundary plane are summarised in Table 5.2.

The most statistically significant collection of data with regard to grain boundary planes has been performed on nickel in a related series of experiments, and so will be reviewed here in some detail. The experiments encompass the ways in which grain boundary energy is minimised where the boundary has rotational freedom, the differences in distribution between boundary plane types on a specimen surface and within the bulk, and one way in which grain boundaries can be engineered to produce lower energy CSLs.

Grain boundaries which traverse a corner of a specimen as shown schematically in Figure 5.6 will rotate so as to minimise locally their energies if the applied driving force is insufficient to unpin the grain boundary edges and instigate grain boundary migration. In the experiments two sorts of rotation and energy minimisation were observed: rotation so as to reduce grain boundary area or rotation towards a 'special' CSL, i.e. a tilt or twist type.[122]

Evidence for the rotations was augmented by the fact that when one surface (say the *XY* plane in Figure 5.6) was open to the annealing environment there was less rotational activity towards lower energy config-

Table 5.2   Summary of experimental data concerning grain boundary planes

Material, specimen	Technique	Format of plane data	Main conclusions	Reference (see appendix)
1. Ni	EBSD	IP scheme	Non-CSLs align with lowest specific area; CSLs rotate mainly to ATGBs	Randle & Dingley (1989)
2. Ni	EBSD	IP scheme	ATGBs, TWGBs and low-index planes well represented.	Randle & Dingley (1990)
3. Ni	EBSD	IP scheme	More tilt and twist CSLs from slow heating than fast heating. CSLs mostly ATGBs.	Randle (1991b)
4. Ni	EBSD	IP scheme	Σ3s STGBs or ATGBs. Σ9s ATGBs. Some low-index planes in non-CSL. Inclination of boundaries correlated with Σ3s.	Randle (1994)
5. Ni	EBSD	IP scheme	Σ3s either ATGBs on 110 zones or STGBs. 211 twin not observed. Irrational planes in non-CSLs.	Randle (1995b)
6. Ni-S	TEM	Mostly qualitative	Relevance of d/a parameter. Most S segregation at boundaries with $d/a > 0.15$.	Bouchet & Priester (1987)
7. Ni-S	TEM	Mostly qualitative	Σ3s: no S segregation to STGBs, some segregation to others.	Bouchet & Priester (1986)
8. Fe-alloy	TEM	IP scheme (some)	7 CSLs, all 'high density'. Fibrous precipitates at these; particle precipitates at other boundaries.	Ainsley et al (1979)
9. NiO	TEM	Graphical	Both CSL and general boundary planes random except no planes near 110.	Dechamps (1991)
10. Cu sheet	SEM	IP scheme	Many boundaries ATGBs. Σ3s & Σ9s often facetted. 2D grains.	Omar (1987); Omar & Mykura (1988)
11. Fe-P	TEM	Graphical	Boundaries often curved. 30% boundaries low-index plane in 1 grain, mostly CSLs.	El M'Rabat & Priester (1988)

Table 5.2 – *Continued*

Material, specimen	Technique	Format of plane data	Main conclusions	Reference (see appendix)
12. fcc steel	TEM	IP scheme	Low-index planes on both sides of boundary. ATGBs more prevalent than STGBs or TWGBs.	Randle (1989a)
13. Si	SEM/ TEM	IP scheme	Interactions of $\Sigma3^n$s. All STGBs, or dissociated into lower-energy segment.	Garg et al (1989)
14. Ni₃Al ribbons	EBSD	IP schemes for Σ3s	Planes in Σ3 nearly all STGBs. STGB Σ3s and Σ1 crack-resistant. 2D grains.	Lin & Pope (1993)
15. bcc Fe-Ni-Cr + P	SEM	Mainly graphical	Low P segregation at all CSLs and non-CSLs with low-index planes.	Ogura et al (1987)
16. Al + Al₂O₃	TEM	Plane indices + Σ	Most planes irrational.	Tweed et al (1984)
17. Nb	X-ray	Plane indices + Σ	11 CSLs, 5 of which Σ1. Most twist or STGB. Facets near triple junctions twist. 2D grains.	Andrejeva (1978)

(IP = interface-plane scheme)

urations than when two sides of the grain boundary surface were open to the annealing environment (both the *XY* and *XZ* surfaces in Figure 5.6).[138] These results are summarised on Figure 5.7 which illustrates the differences in grain boundary geometry when more rotational freedom is permitted by having two surfaces open to the anealing environment rather than one surface. The diagram shows that not only are there more CSLs in the '2-surface' specimen (80%) than in the '1-surface' specimen (50%), but also that in the former there is a dichotomy between boundaries which have rotated to low specific area positions and those that are CSLs, especially with low-index planes. This low-index plane group are almost all Σ3 ATGBs. Figure 5.7 demonstrates clearly that more CSLs, particularly 'special' CSLs with low index planes (ATGBs, STGBs, TWGBs), are produced at the surface of a material, where the boundary has more freedom to rotate, than within the bulk. Further experiments extended this theme with a view to maximising proportions of boundaries having favourable planes and understanding some of the mechanisms underpinning the reactions.

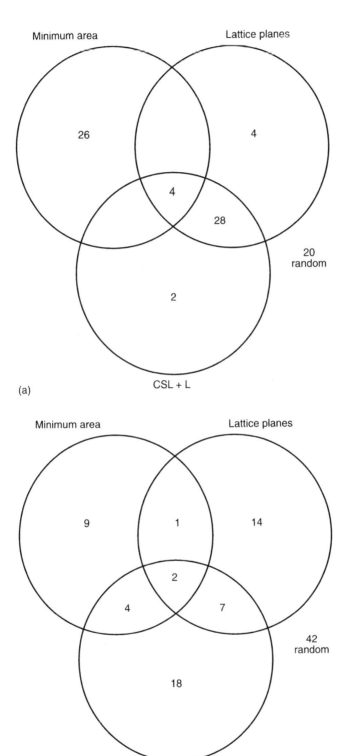

(a)

(b)

Figure 5.7 Proportions of CSL and/or minimum specific area and/or low-index lattice planes grain boundaries for nickel specimens with (a) two surfaces or (b) one surface exposed during annealing.[138]

The next experiment in the series considered two identical nickel specimens which were given a high temperature anneal, 1000°C, for 1 hour. One specimen was raised quickly to holding temperature and subsequently quenched after the anneal (called 'fast') whereas the second specimen was raised slowly to temperature and likewise cooled slowly (called 'slow').[139] The reasoning behind these choices was to investigate if the kinetics of grain boundary migration could be supressed during the slow heating and cooling periods so that boundaries rotated towards lower energy configurations. This mechanism had already been identified during the grain growth incubation period. The boundaries which comprised the sample population were 'corner' boundaries and so had a relatively large amount of rotational freedom compared to boundaries in the bulk.

Figure 5.8 shows that the proportion of CSLs, and in particular Σ3s, is increased dramatically for the slow thermal cycle compared to the fast cycle. Almost all of the CSLs in the former could be classified as tilt or twist boundaries, and these details are shown in Table 5.3. A tolerance of 7° is allowed for each boundary, averaged between both interfacing planes.[48] With regard to ATGBs, if there were no upper limit imposed on the plane indices it would nearly always be possible to find a match with the experimental data. However, planes with very low atomic density, i.e. high

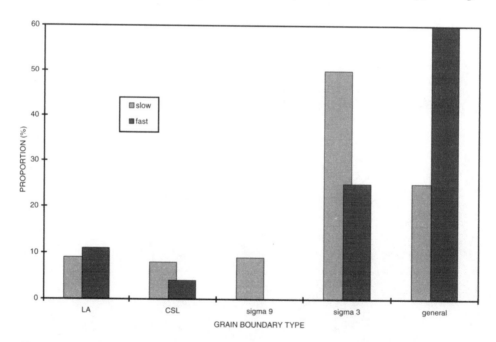

Figure 5.8   Proportions of various boundary types: Σ1, Σ3, Σ9, other CSLs and general boundaries in slow-annealed and fast-annealed nickel specimens.[138]

Miller indices, would have little significance and so the restriction was applied that only boundaries having low-index planes on one or both sides of the boundary, and falling within the average 7° tolerance, are counted as ATGBs.

In the 'slow' specimen 47% of the total boundaries were tilt or twist, compared to 18% in the fast specimen. Table 5.3 shows that most of the boundaries were tilts. In addition to there being more tilt boundaries in the

Table 5.3 Details of grain boundary plane geometries recorded in nickel for 'slow' and 'fast' heat treatment variants (see text)[139]

'SLOW' SPECIMEN

Asymmetric tilts – Σ3s

Σ	V/Vm	Planes
3	0.17	511/111
3	0.13	100/221
3	0.30	100/221
3	0.62	100/221
3	0.28	100/221
3	0.62	100/221
3	0.16	110/411
3	0.08	110/411
3	0.52	110/411
3	0.32	110/411
3	0.16	110/411
3	0.05	122/744
3	0.72	210/542
3	0.18	210/542
3	0.55	110/11,54
3	0.22	211/552
3	0.41	211/721
3	0.22	211/721
3	0.19	211/721
3	0.16	211/552
3	0.30	221/841
3	0.66	221/841
3	0.44	221/744
3	0.12	221/744
3	0.14	221/744
3	0.20	311/755

Symmetrical tilts

Σ	V/Vm	Planes
3	0.17	111/111
3	0.10	111/111
3	0.27	111/111
9	0.39	411/411

Twists

Σ	V/Vm	Planes
3	0.04	210/210
3	0.09	311/311
3	0.45	311/311
9	0.67	322/322
27a	0.21	411/411

Asymmetrical tilts – other than Σ3

Σ	V/Vm	Planes
9	0.74	110/11,54
9	0.57	210/17,10,4
9	0.58	111/13,7,5
9	0.74	111/511
9	0.93	111/511
9	0.25	311/711
17a	0.77	210/30,17,16
19a	0.51	100/18,6,1
19b	0.78	110/25,9,4
11	0.28	311/31,17,9

Table 5.3 – *Continued*

'FAST' SPECIMEN

Asymmetric tilts				Symmetrical tilts		
Σ	V/Vm	Planes		Σ	V/Vm	Planes
3	0.15	221/744		3	0.24	111/111
3	0.69	221/744		3	0.49	111/111
3	0.17	110/411		3	0.33	211/211
3	0.37	511/13,75		3	0.96	211/211
3	0.25	331/11,71				
3	0.48	331/11,71		Twists		
3	0.28	100/744				
3	0.92	111/511		Σ	V/Vm	Planes
3	0.16	311/755				
3	0.55	310/754		3	0.88	210/210
27b	0.83	210/40,37,26		7	0.45	320/320
7	0.75	310/20,93				

slow specimen, most of these boundaries comprise lower index planes than in the fast specimen. In the slow specimen, there is some preference for ATGBs on the 011 zone and, as discussed in the previous subsection, these lie in an energy valley. What is particularly significant is which ATGBs on the 110 zone were observed: 110/411, 100/221 and 111/511. There is an inverse correlation between the frequency of occurrence of the ATGBs and their energy, as shown on Figure 5.9. But furthermore, there are many other 110 ATGBs which were not observed even though they had lower energies (see Figure 5.4). It is suggested that boundaries having *one low index plane,* and hence a high value of d_{eff}, were selected preferentially when boundaries rotated. The absolute boundary energy must still play a role or else 111/511 planes would have occurred more frequently and 211/211 might have been observed. Probably there is a compromise effect between boundary configurations which are attainable within the geometrical constraints of the polycrystal and the selection of low energy planes.

The final two experiments in the series to investigate the manipulability of grain boundary planes in nickel considered the differences between boundaries abutting one surface (but not two surfaces) and those within the bulk of a specimen. Figure 5.10 shows the distribution of grain boundary inclinations, expressed as the angle between the grain boundary surface and the specimen surface normal, both for boundaries which abutted the specimen surface during annealing and those within the bulk. The plot shows that

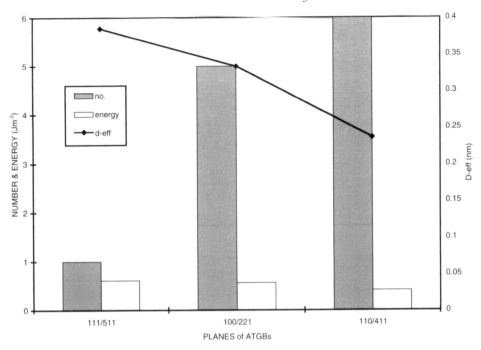

Figure 5.9   Occurrence of three types of Σ3 ATGBs having low index planes in a nickel specimen, as a function of energy and d_{eff}. The data points for d_{eff} are joined as a guide to the eye.

there is a tendency for boundaries at the surface to become upright. However, those boundaries at the surface with high inclination angles were almost all Σ3s – a similar observation to some of the grain boundary 'corner' experiments.[122]

Measurement of boundary plane crystallography in the interior of the specimen revealed that all CSLs except Σ3s had irrational boundary planes.[140] Half of the Σ3s were ATGBs on the 1 1 0 zone. As in previous experiments on nickel, the 2 1 1/2 1 1 STGB was not observed. Figure 5.11 shows an example of boundaries in nickel whose plane indices have been determined. For the boundaries which abutted the specimen surface during annealing, a larger proportion of 'special' boundaries was observed than for boundaries in the bulk of the specimen, including more Σ3 ATGBs on the 1 1 0 zone, more ATGBs for other CSLs and more boundaries (CSLs and non-CSLs) having low-index planes in both grains.[141]

In summary, this series of experiments to investigate grain boundary plane geometry in pure nickel has concluded that grain boundaries can be engineered, by judicious use of heat treatments, to produce a high level of 'special' boundaries with regard to their planes. The enhancement value of

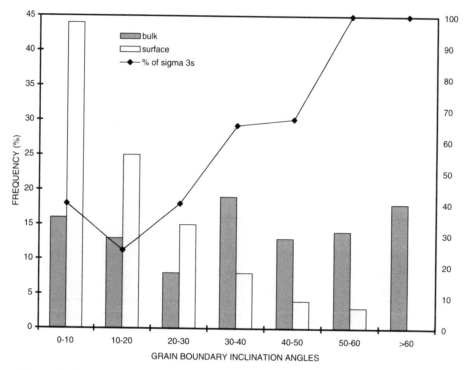

Figure 5.10   Frequency distribution of inclination angles (angle between the grain boundary plane and the surface normal) for two sample populations of boundaries in pure nickel: one population is at an annealed surface whereas the other population is within the bulk of the specimen. The proportion of $\Sigma 3$s for the surface specimen is also included, showing that there is an inverse correlation between $\Sigma 3$ proportions and inclination angle.[140,141] The data points for proportion of $\Sigma 3$s are joined as a guide to the eye only.

these grain boundary engineering processes can be assessed by comparison to the baseline population in the bulk, which shows lower proportions of special boundaries. Although the database concerning distributions of CSLs is increasing rapidly (see Figure 6.9), the amount of data which also reports boundary planes in the CSL for polycrystals is very small by comparison – Table 5.2 summarises the main reports. In this light the data in Table 5.2 provide valuable statistics on distributions of boundary planes, and the following is a summary of the most significant points:

• Frequently grain boundary planes are irrational, even when a boundary is a low-Σ CSLs (numbers 1–6, 10, 16).
• In nickel and copper ATGBs occur far more frequently than STGBs (numbers 1–7, 10).
• STGBs and other special CSLs occur more frequently where the bound-

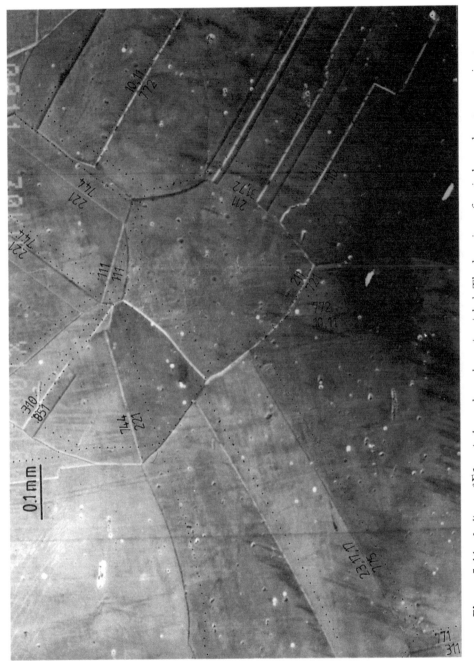

Figure 5.11   Indices of Σ3 grain boundary planes in nickel. The location of grain boundary traces prior to serial sectioning is indicated by a dotted line.

aries possess an amount of thermally activated rotational freedom, e.g. in a two-dimensional Ni$_3$Al polycrystal (no. 14), Cu polycrystal (no. 10) and Nb polycrystal (no. 17), or where boundaries traversed the 'corner' of a nickel specimen (numbers 1–3). Furthermore, boundaries on a specimen surface are 'pseudo two-dimensional', that is, they have more rotational freedom during annealing than those within the bulk (numbers 4, 5).

- STGBs predominated in a Ni$_3$Al specimen (no. 14) and in a polysilicon specimen (no. 13). In the latter case ATGBs dissociated to STGBs as illustrated in Figure 5.12 which shows the dissociation of an asymmetric portion of a $\Sigma 27$ into $\Sigma 9$ and $\Sigma 3$ boundaries.[38] Such dissociation reactions are discussed further in Section 5.2.2. The differences in behaviour in these two materials shows that there are variations in relative energies for boundaries having different bonding and electronic structures.
- Grain boundary properties – energy (numbers 1, 2, 13), mobility (numbers 3, 11), segregation (numbers 6, 7, 15), precipitation (no. 8), cracking (no. 14) – all show a correlation between low values and 'special' CSLs, i.e. those which do not have random boundary planes.
- There is a body of results from several different materials (numbers 4, 11, 12, 15) which demonstrate that low-index grain boundary planes occur frequently, either in one grain or, more significantly, in both grains. Although this configuration is not part of the CSL scheme, it may be of importance to grain boundary behaviour, especially when it is considered

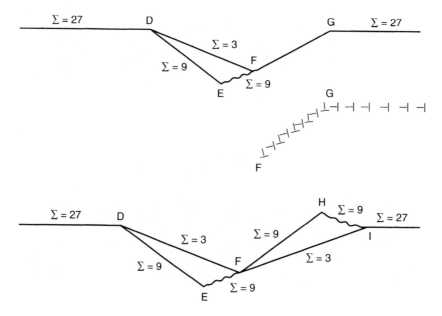

Figure 5.12   Schematic illustration of how $\Sigma 27$ boundaries dissociate in polysilicon.[38]

that low-index planes interfacing boundaries have been observed in HREM.

- Accommodation effects near triple junctions give lower density CSLs (no. 17).

To conclude this section, only 10% of CSL-based experiments include the boundary plane as a parameter. These experiments have shown that a CSL boundary does not necessarily have 'special' planes. Low-index planes may also be important.

5.2 GRAIN BOUNDARY AND JUNCTION CONNECTIVITY

5.2.1 Introduction

In Section 2.2.3 the geometrical rules which govern the relationship between three CSLs which meet at a junction were defined. In the present section we will show that these rules have a significant effect on the CSL distribution, particularly when Σ3 boundaries are involved. Consideration of the interactions at triple junctions is the first step towards analysing the collective behaviour of polycrystals, which includes the network characteristics, or *connectivity*, of grain boundaries. This will be discussed briefly in Section 5.2.3.

5.2.2 Grain junctions

A CSL analysis of grain boundaries in polycrystals can be extended to grain junctions by counting the proportion of CSLs in each triple junction (i.e. 1, 2 or 3 CSLs) and analysing the identity of those CSLs.[90,142] For grain boundaries which are very close to exact CSLs, the CSL geometry can be predicted from the rules given in Section 2.2.3. Where deviation from the exact CSL configuration of a grain boundary at a grain junction is near the upper angular limit, the other boundaries at the junction may not fall within a CSL range, and so CSL multiplication is not perpetuated.[124]

From a grain boundary engineering point of view, triple junctions having two or more special boundaries (e.g. Σ3s, Σ1s) are beneficial elements of microstructure because interfacial degradation is suppressed at these junctions. Such junctions may also be stable against grain growth, and have been called 'secure' junctions. Figure 5.13 illustrates schematically the effect of secure junctions on the microstructure.[126] The theoretical limit to the proportion of Σ3s in a microstructure is 2/3, i.e. two at every junction.[143] The third boundaries at each junction would be Σ9 or Σ1. Twinning has already been exploited in grain boundary engineering (see Chapter 1 and

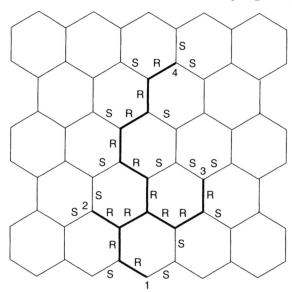

Figure 5.13   Schematic illustration of a crack propagating from 1 through a grain boundary network of special (S) and random (R) boundaries. The crack arrests when it encounters 'secure' triple junctions (labelled 2, 3 and 4) i.e. those having two special boundaries.[126]

Section 6.1.4). The presence of two low-energy $\Sigma 3$ or $\Sigma 1$ boundaries (or one of each) at a triple junction is beneficial to reduction of intergranular degradation since the network effect for transport phenomena is considerably reduced when only one boundary at a junction supports easy diffusion. With regard to grain boundary engineering, it has been demonstrated that the connectivity parameters of low energy boundaries can be improved in nickel by processing, such that proportions of secure junctions have been increased from 37% to 61%.[12]

The $\Sigma 3^n$ family gives rise to a whole hierarchy of interactions at grain junctions, both in terms of grain boundary dissociations, as illustrated in Figure 5.12, and perpetuation of CSLs by interactions at junctions. Both of these reactions have been commonly observed in several low-stacking fault energy systems for the $\Sigma 3–\Sigma 3–\Sigma 9$ interaction given in equation 2.5.[38,113,144,145] The case for $\Sigma 3^n$ where n = 3, 4, 5 ($\Sigma 27$, $\Sigma 81$ and $\Sigma 243$) is more complex. For example, it has been observed in a Cu–6Si at.% alloy that $\Sigma 27a$ boundaries did not dissociate to give $\Sigma 3$ and $\Sigma 9$ boundaries, but rather gave $\Sigma 3$ and $\Sigma 81d$ boundaries. Similarly, $\Sigma 81d$ boundaries always dissociated to $\Sigma 3$ and $\Sigma 243a$ rather than to $\Sigma 3$ and $\Sigma 27$. These results were explained in terms of both boundary symmetry to retain the 1 1 1 coherent plane as the twin boundary and the kinetics of dissociation, rather than simply low energy.[146] However, these reactions may not be duplicated in other materials; for example in silicon $\Sigma 27a$ boundaries gave rise to a $\Sigma 3$ and $\Sigma 9$ rather than a $\Sigma 81$.[38] These discrepancies between materials again highlight the purely geometrical nature of the CSL model.

It should also be noted that the $\Sigma 243a–\Sigma 3–\Sigma 81d$ grain junction configuration, which has been identified in an austenitic steel, could also have been interpreted as a $\Sigma 1–\Sigma 3–\Sigma 3$ trio since the misorientations for $\Sigma 243a$ ($7.4°/110$) and $\Sigma 81d$ ($60.4°/443$) are within the limits of $\Sigma 1$ and $\Sigma 3$ respectively, according to the Brandon criterion. The authors decided on the former criterion from examination of the geometry of adjacent grain junctions and morphological features of the boundary.[147] Again, the geometrical limitation of the CSL is apparent.

So far only grain junctions comprising three boundaries – triple junctions – have been mentioned. Although these are by far the most common type, it is possible for four or even five boundaries, usually from the $\Sigma 3''$ family to co-exist in a stable configuration at a junction. Figure 5.14 shows four $\Sigma 3$s and one $\Sigma 81$ boundary at a junction in nickel.[73,148]

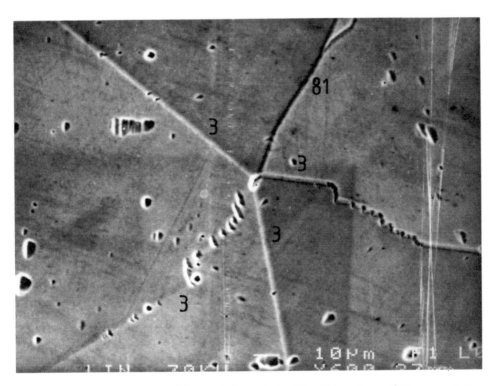

Figure 5.14 Conjunction of five grain boundaries (four $\Sigma 3$s and one $\Sigma 81$) at a node in pure nickel.[21]

Computer modelling is a convenient tool for studying CSLs at grain junctions, and is often used in conjunction with data from real micro-structures. For example, it has been shown in simulations that 110 fibre texture increases the proportion of triple junctions having two Σ3s from 3% to 19%.[147] Other simulations have shown that the fraction of junctions having two or more CSLs is increased more for a 100 or 111 fibre texture than for a 110 fibre texture.[90] Another experiment, involving both simulation and measurements from microstructures, shows that in pure aluminium there is a correlation between the degree of grain size uniformity and fraction of random (i.e. Σ >29) boundaries at a grain junction, and also between the heat treatment temperature and the proportion of random boundaries at a junction as indicated on Figure 5.15.[142]

In summary, this subsection has shown that an important aspect of the role of the CSL in grain boundary engineering is the adjustment in the proportion of certain CSLs, particularly the Σ3 family, caused by interactions at triple junctions.

5.2.3 Grain boundary connectivity

Consideration of the geometrical relationships at triple junctions is the first

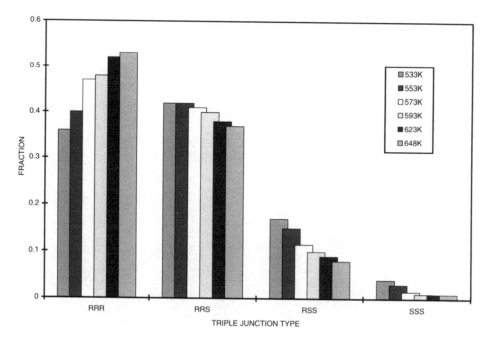

Figure 5.15   Triple junction compositions in terms of special (S) or random (R) boundaries in aluminium annealed at temperatures ranging from 533 K to 648 K.[142]

step towards combining the topological and orientational parameters of the polycrystal. Such an approach is important for grain boundary engineering because the way in which grain boundary properties affect the whole polycrystalline aggregate depend on the *distribution* of special boundaries throughout the network. Preliminary experiments have shown that grains of particular orientation may be spatially distributed in a non-random manner – this is referred to as orientation 'clustering' or 'coherence'.[82] In consequence, grain misorientations are similarly non-random. Other factors may also exist which exert local effects on misorientations and hence CSLs, such as chemistry imhomogeneities and external geometrical influences. An example of the latter is that the distribution of CSLs has been observed to be different at surface grains to non-surface grains in rolled interstitial-free steel sheet (Figure 5.16).[149]

To date, there have been relatively few attempts to incorporate connectivity into grain boundary engineering. Some have been already mentioned in Section 3.2, i.e. the Intercrystalline Structure Distribution Function[105] and the Grain Boundary Correlation Number.[93] Other attempts have used a computer simulation methodology based on percolation theory

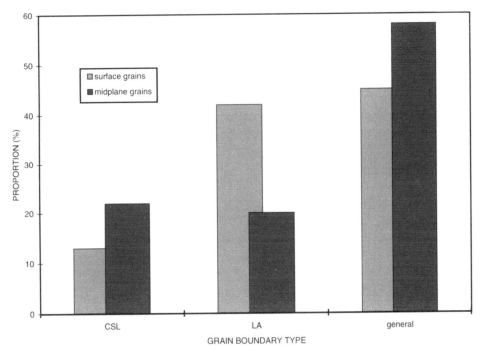

Figure 5.16 Proportions of grain boundary types in interstitial-free steel sheet for a sample population at the sheet surface and another from midplane grains.[150]

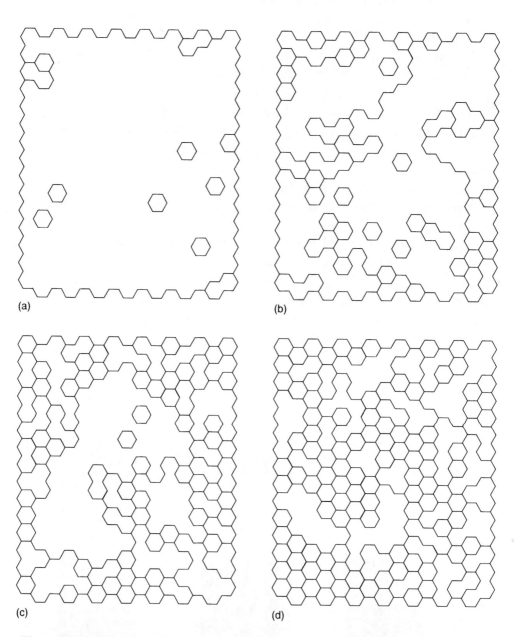

Figure 5.17   Computer simulations showing the evolution of 'effective' grain sizes when the proportion of $\Sigma 1$ boundaries plus unconnected high angle boundaries is removed from the microstructure. The simulations are for grains spread (a) 20°, (b) 25°, (c) 30° and (d) 35° from exact cube texture.

to characterise the cluster properties of grains.[72] As yet, this work uses a simple binary division based on the properties on Σ1s and random high angle boundaries. It was suggested that the cluster size, i.e. a group of grains joined by Σ1s, was more important to some material properties than the grain size.

The research into cluster properties has stimulated further work involving simulation of near cube, and other, textures.[112] The near cube texture has been analysed in terms of Σ1s, since no other CSLs were present in proportions greater than that expected for random generation. Figure 5.17 shows an 'effective grain size' when Σ1 and disconnected random boundaries are discounted. The rationale for excluding Σ1s is that they have markedly different properties to high angle boundaries, and so transport phenomena etc would affect only the high angle boundaries. A similar effect has been observed in aluminium thin films for interconnects, whose unusually long lifetime has been explained by the presence of chain-like and equiaxed 'colonies' of grain having special boundaries between grains within colonies.[66]

6. *The Role of the Coincidence Site Lattice in Grain Boundary Engineering*

6.1 SELECTED AREAS OF STUDY

6.1.1 Introduction

In this Section three areas which represent significant application of the CSL model will be appraised in more detail. These are orientation relationships during recrystallisation, anomalous grain growth in iron-silicon alloys and twinning.

6.1.2 Orientation relationships during recrystallisation

After nucleation, recrystallisation is controlled solely by grain boundary migration. It follows then that in order to understand the mechanisms of recrystallisation, it is essential to understand which grain boundaries can migrate most readily. To this end, workers have found it convenient to categorise orientation relationships across the recrystallisation interface by Σ-value. The most important conclusions obtained in recent years, mostly on aluminium[151,152] and copper alloys[153,154] indicate that

- There are differences in behaviour between materials and between alloys of the same material. For example, a boundary having a misorientation of 40°/1 1 1, which is near Σ7, is credited with particularly high mobility and yet is only prevalent during recrystallisation in aluminium.
- New orientations were observed to be created by repeated twinning[15] even in high stacking fault energy materials such as aluminium. One of four twin variants is selected, and at present the criteria for selection is still unclear.
- Orientation relationships may be in the form of high Σ CSLs.[155]
- Although it is postulated that CSLs augment the recrystallisation process it is unclear which is the important parameter: low energy or high mobility.

A CSL is generated at an interface as a direct consequence of the relationship, at the boundary, between the texture of the two neighbouring grains. Hence, for example, a typical rolling texture, $(0\,1\,1)(2\,1\,\bar{1})$, and recrystallisation texture, $(3\,\bar{1}\,1)[1\,1\,\bar{2}]$, for austenitic steels is related by a misorientation which is <1° from Σ29.

All CSLs can be related to the texture of neighbouring grains in a similar way. A pertinent question is, *to what extent can measured textures and measured CSLs be idealised and still be physically valid?* Orientation relationships which are quoted in the literature are related to the nearest CSL, which may be up to 10° away[115] – certainly outside the Brandon criterion.[46] Furthermore, the existence of considerable lattice curvature in a heavily deformed material means that the misorientation, and therefore the closeness to a particular CSL, varies over the whole grain boundary area. Figure 6.1 illustrates this variation with data from 40% deformed aluminium, which shows the variation in orientation along a grain boundary on a micrometre scale.[156] To quote an 'average CSL' is not justified by the actual relationship at the boundary plane, which is changing. This fluctuation will in turn affect any special properties at the boundary.

The consequence of the lack of exactness in many CSLs ascribed to orientation relationships across recrystallisation interfaces is that they cannot be linked rigorously to special properties. Instead they are restricted to a classification role, and used as a shorthand way to highlight differences between data sets. An exception is the twinning phenomenon, where it has been established that Σ3s play a major role in the onset of the growth phase of recrystallisation.

6.1.3 Secondary recrystallisation in iron-silicon alloys

There is a direct technological interest, for Fe–3%Si transformer steels, in the growth of very large grains having the so-called 'Goss texture', (1 1 0)[0 0 1], since this is the condition for development of optimum magnetic proper-

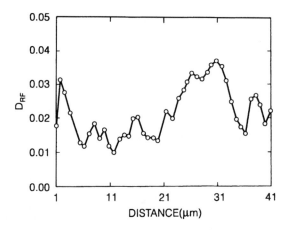

Figure 6.1   Fluctuations in orientation along a grain boundary in pure aluminium deformed 30%, as a function of the parameter D_{RF}, which is a compressed Rodrigues vector. The orientations were measured at 1 μm intervals.[156] (Randle et al, 1996.)

ties.[157] Such exaggerated grain growth requires a few boundaries to be mobile while the rest are stationary. Study of the mechanism of this secondary recrystallisation has generated much research, which necessarily focusses on the role of grain boundaries, since it is these which actually implement anomalous grain growth. Figure 6.2 shows the early stages of secondary recrystallisation in Fe–3%Si.

A central tenet which governs the onset of secondary recrystallisation is that only *some* grain boundaries are mobile thus enabling selected grains to grow. The important questions, insofar as understanding the phenomenon of anomalous grain growth is concerned, is *which boundaries*? and *why*?. The CSL model was a natural choice as an analytical tool to categorise grain boundary structure in these alloys since well-defined textures are involved after both primary and secondary recrystallisation, from which data CSL statistics can be calculated readily.[158]

The most notable research work on the role of CSLs in secondary recrystallisation of Fe–3%Si has been the 'Simulation by Hypothetical nucleus', or SH, method which is used to predict whether or not secondary

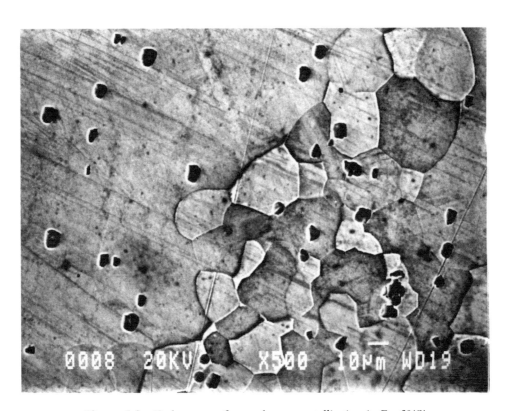

Figure 6.2   Early stages of secondary recrystallisation in Fe–3%Si.

recrystallisation will occur.[159] The crucial initial premise in the SH model is that it is necessary to know the identity of mobile boundaries present at the onset of grain growth, and also all through the grain growth process, since grains which are in contact with each other will change as grain growth proceeds. Measurement of misorientations between large/small grains *after* grain growth does not provide insight into which boundaries were able to trigger grain growth, i.e. were most mobile, in the first place – those initial nucleus grains are now inaccessible experimentally. To overcome this problem a *hypothetical secondary recrystallisation nucleus* is specified, which is allocated to an orientation known to exist as the final grain growth texture. For the case of Fe–3%Si transformer steels, this is (1 1 0)[0 0 1].

The SH procedure is applied as follows. Orientation measurements are made of several hundred grains in primary recrystallised material, either using a single orientation measurement method or X-ray texture. For the latter, an analytical procedure is used to obtain the proportion of grains clustered about different orientations. The misorientations between all the experimentally determined orientations and the hypothetical nucleus orientation are then calculated and analysed according to the CSL distribution. The frequency of CSLs thus obtained is I_C. It turns out that there are I_C peaks for CSLs oriented on the 1 1 0 axis, $\Sigma 9$ in particular, and the assumption is made that these boundaries constitute the mobile types. The parameter P_{CN} is then calculated as the product of I_C and the intensity of nucleus orientations actually present in the material, I_N. A critical value of P_{CN} then needs to be exceeded for secondary recrystallisation to take place.

Figure 6.3 illustrates the relationship between P_{CN} and deviation from the exact Goss orientation for three specimens. For only one of these did the P_{CN} value at the exact Goss orientation exceed the P_{CN} threshold, which is 45 in this case.[159] Subsequent annealing did indeed show that only the specimen with a high P_{CN} developed the desired Goss texture after grain growth. This work demonstrates that the SH method, based on the distribution of CSLs between the nucleus and other orientations, can be used to predict the occurrence of specific secondary recrystallisation textures. The SH method can be applied to predict other secondary recrystallisation parameters. For example, cross rolling can produce a (1 0 0)[0 0 1] secondary recrystallisation texture in Fe–3%Si, which is related by $\Sigma 7$ to the primary texture.[160] Experiments have established that P_{CN} for $\Sigma 7$ is highest in the midplane of the rolled sheet.

A key aspect to the success of the SH predictive model is that strong and fairly simple textures are involved. This gives rise to two effects: most grain pairs are related by either simple orientation relationships, often CSLs, or a

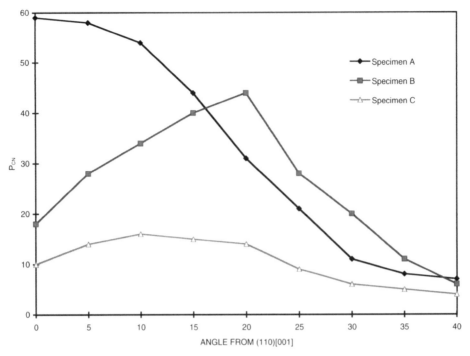

Figure 6.3 Angle from the Goss orientation, $(110)[001]$, for three Fe–3%Si specimens as a function of the P_{CN} value. Only specimen A, having a high P_{CN}, subsequently developed secondary recrystallisation.[159]

$\Sigma 1$ relationship because they are near the same orientation. For example, the Goss orientation and primary recrystallisation texture are related as follows:

$$(1\bar{1}0)[001] : (111)[11\bar{2}] = 35.3°/110$$

This misorientation is less than 3° from $\Sigma 9$. Data from single crystal Fe–3%Si, rolled and recrystallised then subsequently given a grain growth anneal show that the strong texture present before grain growth means that more than 80% $\Sigma 1$ boundaries are present. A $\{120\}<001>$ texture predominated after grain growth; the proportion of $\Sigma 1$s had reduced to 20% and CSLs with a 110 misorientation axis were present.[158]

It is clear then, from a fairly substantial body of evidence, that CSL boundaries are instrumental in accomplishing secondary recrystallisation in transformer steels to give a strong Goss texture, although the mechanisms are not yet fully elucidated. Two reasons why CSLs are so active are:

• There is a high statistical occurrence of particular CSLs because of the simple relationship between orientations;
• The presence of strong textures dictates that many $\Sigma 1$ boundaries are

present; these are more immobile than other CSLs and so effectively pin the microstructure.

Fine precipitates have an additional pinning effect on the microstructure. It is probable that they will be most effective at random high angle boundaries, thus leaving low-Σ CSLs as the only boundaries capable of migration. Grain growth control in this manner can be considered to be a form of grain boundary engineering, since it is necessary to obtain the correct proportions and distributions of immobile or pinned boundaries and CSLs.

6.1.4 Twinning

Throughout this book many examples have been quoted of the occurrence of particular CSL boundaries and whether they do or do not exhibit special properties. The one CSL which *always* exhibits special properties, except under the most severe environmental conditions, is the $\Sigma 3$ coherent twin. It is therefore beneficial to overall intergranular degradation properties to have a high surface area of $\Sigma 3$s relative to the total grain boundary surface area, and indeed this has formed the basis of Grain Boundary Engineered (GBE[TM]) Alloy 600. To recap briefly, the following aspects of $\Sigma 3$ boundaries have been addressed in this book:

- The capacity for twin formation correlates with low stacking fault energy. For example austenitic steels, nickel alloys and copper alloys twin readily.[123]
- $\Sigma 3$ boundaries interact with other boundaries at grain junctions. Where a high proportion of $\Sigma 3$s is present, families of $\Sigma 3^n$ boundaries can be generated.[38,146]
- The $\Sigma 3$ CSL has the highest multiplicity of all CSLs, and resides on the apex of the fundamental zone or asymmetric domain.[107]
- The 111 STGB in the $\Sigma 3$ system, the coherent twin, is associated with a deep energy cusp, immobility and resistance to intergranular degradation.[51]
- $\Sigma 3$ boundaries can also be characterised by planes other than 111. These $\Sigma 3$s do not have energy as low as the 111 STGB, yet still have lower energy than general high angle boundaries.[134]
- The onset of recrystallisation is often facilitated by twins, even in high stacking fault energy materials.[15]

At present twinning and its ramifications are the most effective route applied for grain boundary engineering. The general requirement is to increase the proportion of twins and twin-related boundaries while maintaining a small grain size. In other words, the proportion of total grain interface *area* which is twins or other special boundaries should be maximised. Grain boundary engineered Alloy 600 has a grain size of less than 20 μm.[12]

Since it is established that increasing Σ3s is the key to one grain boundary engineering route, the question arises: *what is the maximum frequency of twins which can be accommodated in a microstructure?* Geometric considerations have shown that this figure is 2/3, i.e. a maximum of two out of three interfaces could be a Σ3. The situation at triple junctions would then be as shown schematically in Figure 6.4. From the association rules of CSLs given in Section 2.2.3 the third boundary would be either a Σ1 or a Σ9. Such a configuration forms what has been termed a *'twin-limited microstructure.'*[143] As far as energy – and therefore properties – are concerned, in such a microstructure the 'energy hierarchy' of the constituent boundaries would increase as follows:

• Σ3s on 1 1 1 planes
• Σ1 (depending on the magnitude of its misorientation)
• Σ3s on the 1 1 0 zone as discussed in Section 5.1
• Σ3s and Σ9s which are STGBs (except 1 1 1), ATGBs and TWGBs
• Σ3s and Σ9s with irrational boundary planes

In a twin-limited microstructure the boundaries in the last category would not have special properties. However, the other four groups would collectively form the majority of boundaries producing grain junctions at which

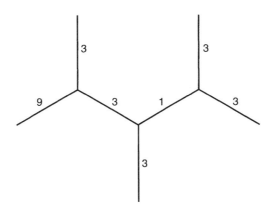

Figure 6.4 Schematic illustration of how a 'twin-limited' microstructure could evolve resulting in regions containing Σ3–3–9 or Σ3–3–1 combinations at grain junctions. Boundaries are labelled with their Σ-value.

at least two out of three boundaries are special, thus fulfilling the condition for a 'secure' triple junction, as described in Section 5.2, which can effectively arrest intergranular degradation.[126]

Although individual incidences or small groups of $\Sigma3/\Sigma3/\Sigma9$ or $\Sigma3/\Sigma3/\Sigma1$ have been observed experimentally, a twin-limited microstructure consisting *entirely* of $\Sigma3$, $\Sigma9$, $\Sigma1$ is an idealised state which has not been achieved in practice. Where enhanced proportions of $\Sigma3$s have been observed these have tended to be accompanied by proportions of $\Sigma9$s which are higher than that expected for random generation, thus indicating some interactions between twins.[145,161,162] Figure 6.5, which comprises data from 46 reported instances in the literature of both $\Sigma3$ and $\Sigma9$ proportions, illustrates this trend. In addition to $\Sigma9$ other $\Sigma3^n$ boundaries can be generated by boundary interactions, as discussed in Section 5.2. Two points to recall in this connection are that $\Sigma81d$ ($\Sigma3^4$) has a misorientation which is near to $\Sigma3$ and $\Sigma243a$ ($\Sigma3^5$) is a low angle boundary (Section 5.2.2). Hence a sustained amount of multiple twinning and interactions between the products can give rise to *regeneration* of near-$\Sigma3$s and $\Sigma1$s. These 'pseudo $\Sigma3$s', i.e. renamed $\Sigma81$s, will not have boundary planes on 111; however

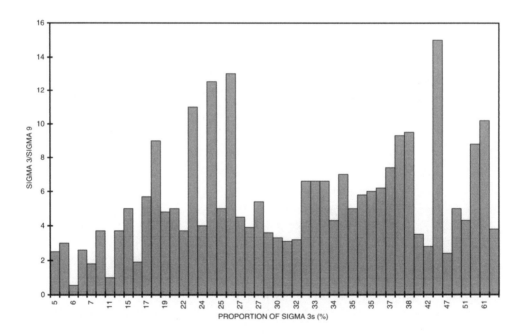

Figure 6.5 Graph showing proportions of $\Sigma3$s, in ascending order, as a function of the ratio $\Sigma3/\Sigma9$ where both these proportions are quoted. There is no consistent relationship between these fractions.

under favourable thermomechanical treatment conditions it is possible for boundary planes and misorientations to rotate into lower energy positions as demonstrated in Section 5.1.5.

Grain boundary engineering in Alloy 600 is a commercial process and, as such, details of the processing route used to obtain enhanced proportions of special boundaries, which were illustrated on Figure 1.2, have not been published. Although for this case the mechanisms for production of heightened twin fractions are not available, a series of experiements on pure nickel has aimed to elucidate some of the fundamental aspects of twinning in 99.999% and 99.5% nickel. Pure nickel has a coherent twin boundary energy of 43 mJm^{-2} (obtained from the stacking fault energy) which means that the absolute numbers of annealing twins which can be generated are necessarily lower than for many austenitic steels or nickel alloys where the coherent twin boundary energy is typically <20 mJm^{-2}.[50] However, *changes* in the twin proportions can be effectively monitored and analysed whatever the absolute twin numbers.

The essential condition which drives the formation of new annealing twins is that the resultant energy of the twin plus two new boundaries must be less than the original boundary configuration which they replace.[163] However, this criterion is not straightforward because the twin is in contact with many other grain boundaries in addition to those viewed in a single polished section, and therefore any energy minimisation must apply to the system as a whole rather than to individual boundaries.[164] It has generally been observed to be the case that the twin fraction changes during grain growth, usually to higher twin proportions.[165] These increases are not usually very dramatic, and furthermore exploitation of twinning for grain boundary engineering often requires that the grain size remains fairly constant. Processes other than grain growth, then, are required to induce twinning.

At this point it is important to emphasise the difference between Σ3 *grain boundaries* and Σ3 *twins*. Σ3 boundaries are grain boundaries which have a misorientation within 8.6° of 60°/1 1 1. Certain textures will introduce them into the microstructure, for instance textures which are common in fcc materials such as brass, copper and Σ have variants which are Σ3-related. For example in the case of the brass texture the components (1 1 0)[1 $\bar{1}$ 2] and (1 1 0)[1 $\bar{1}$ $\bar{2}$] are twin related.[166] A strong 1 1 0 fibre texture also increases the frequency of Σ3s and other boundaries misorientated on 1 1 0. These Σ3 grain boundaries will only have special properties if the boundary plane has certain rational indices. A Σ3 twin, on the other hand, has been generated by a 'twinning event' i.e. a 180° rotation of the lattice on the close-packed plane of the CSL to give a coherent twin plus regions of incoherent twin facets. The incoherent twin is not usually on the 2 1 1 plane; experimental evidence for this was shown in Section 5.1.5.

Between the two extreme cases of non-special $\Sigma 3$ boundaries and 1 1 1 twins there are a range of $\Sigma 3$s, having several possible origins, which have lower energies than average boundaries and so will have special properties. The question arises of how to distinguish these types. Ideally if the indices of the boundary plane are known, the specialness criterion is self-evident. Figure 6.6 shows the correlation between the relative deviation of the misorientation from exact matching, expressed as v/v_m (see Section 2.2.4), and the type of grain boundary planes. The majority of special planes, particularly 1 1 1 planes, are characterised by $v/v_m <0.2$, and so this has been used as a guide to distinguish special/non special $\Sigma 3$s from misorientation data alone.[167] Furthermore special $\Sigma 3$s usually have a straight appearance in the microstructure. An 86% correlation was found between $\Sigma 3$s having $v/v_m <0.2$ and a planar morphology, suggesting that this is a suitable working

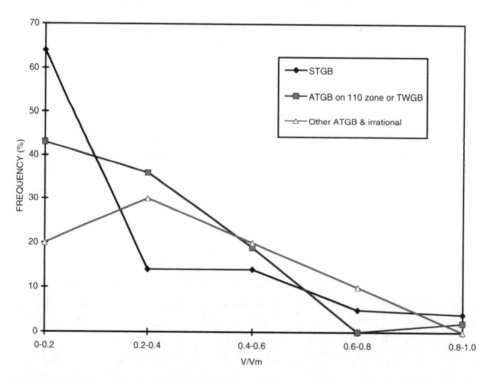

Figure 6.6   Proportions of different $\Sigma 3$ types in a pure nickel specimen as a function of v/v_m illustrating that many, but not all, $\Sigma 3$s having $v/v_m <0.2$ are STGBs or other low volume configurations.[140]

threshold for use to distinguish special CSLs.[168] Clearly it can be an oversimplification to determine only the proportion of Σ3s in a material whereas the simple step of including their proximity to exact matching is much more revealing.

Studies of factors which influence increases of Σ3 boundaries in nickel have shown that:

- After anomalous grain growth was induced in superpure nickel by annealing in vacuum at 0.68 T_m, the special Σ3s were sited in the small-grained regions, rather than between small/large grains or associated with the anomalous grains. The statistics of this observation are shown on Figure 6.7. This is evidence that twins are generated when other mechanisms for grain boundary energy reduction (grain growth, grain rotations) are restricted.[165]
- When superpure nickel is strained, given a low temperature (500°C) preliminary anneal in air followed by 'ageing' at 750°C , a ten-fold increase in Σ3s results. This is attributed to a drag effect on the boundaries from oxygen ingress resulting in boundary migration restriction.[126]
- Building on the previously mentioned results,[126] the proportion of Σ3s, and more especially the proportion of them having v/v_m <0.2, is affected by the precise thermomechanical treatment. Small changes in the schedule can influence the balance of special/non-special Σ3s, as shown on Figure 6.8,

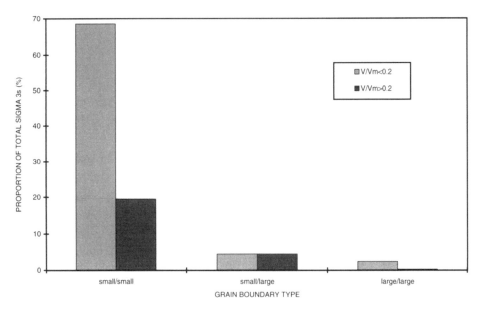

Figure 6.7   Breakdown of CSLs according to v/v_m and size of interfacing grains in an area of anomalous grain growth in a pure nickel specimen annealed at 0.68 T_m.[165]

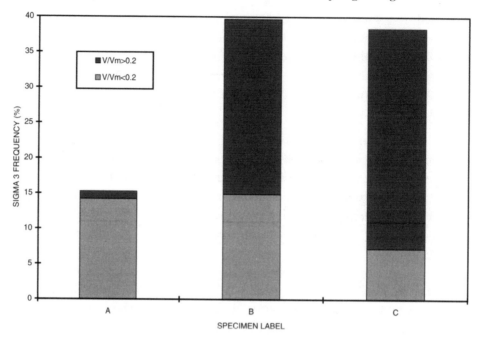

Figure 6.8   Distribution of $\Sigma 3$s according to v/v_m for three heat treatment variants in
pure nickel.

by either generating more new twins or 'fine tuning' those $\Sigma 3$s already
present so that (in the absence of grain growth) their planes can equilibrate
to lower v/v_m values.[168]

• The *rate* of grain boundary migration appears to be critical. Fast moving
boundaries do not nucleate twins as readily as slower moving boundaries.[165]

In summary, the experimental evidence such as that in Figure 1.2 and
from many other sources is unequivocal in showing that twinning enhances
resistance to many kinds of intergranular degradation and therefore it is
desirable to increase the proportion of $\Sigma 3$ and $\Sigma 3''$ boundaries in a material.
This subsection has highlighted and brought together some of the complexities of $\Sigma 3$ generation, interactions and characteristics. The mechanisms and
hence the optimum routes for $\Sigma 3$ generation are currently being explored,
and it seems likely that a balance between available driving force and grain
boundary migration rate are key elements in maximising $\Sigma 3$ fractions.

6.2 CONCLUDING REMARKS

The aim of this book has been to evaluate critically the role of the CSL
model in modern grain boundary investigations, and ultimately to assess its

contribution to the emerging field of 'grain boundary design'. Throughout the book the theme has been how the CSL model has been used as an analysis tool. This has been demonstrated by statistial analysis of experimental evidence. It is apparent that the CSL model features frequently in grain boundary research. Figure 6.9 shows how the number of investigations using CSLs in cubic polycrystals has changed in the last two decades, and it is seen that the numbers increase greatly from 1987 onwards. Although this trend is partly reflecting improved experimental techniques, it also establishes that there is great interest from the scientific community in exploiting the CSL model as an indicator of special properties.

Without doubt work in recent years has shown that **the proportion of CSLs in a polycrystalline material can be manipulated**. The *starting points* which dictate the proportion of CSLs already present relate principally to the interrelationship between microtexture, chemistry (including stacking fault energy and boundary pinning) and grain size distribution. The *tools* for altering the CSL profiles are the introduction of driving forces for migration (or sliding, rotation, nucleation of twins, etc) which act upon grain boundaries, heat treatments to allow the boundary network to move towards equilibration and subtle alterations in local chemistry, e.g. selective use of annealing environments. The desired *effect* of these treatments is to increase the proportion of CSLs as a route to increasing the proportion of grain boundary area which has special properties.

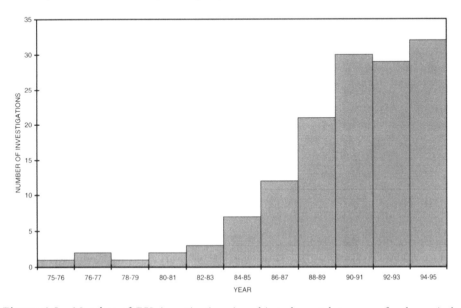

Figure 6.9   Number of CSL investigations in cubic polycrystals per year for the period 1975–1995.

Having established that the CSL population is metastable and can be modified, the second question which can now be answered is: does the experimental evidence support a link between CSLs and special properties? Notable demonstrations of such a connection in polycrystals have been quite specific to particular CSLs and material conditions. In other words, the idea that *all* low-Σ CSLs are beneficial is not borne out in practice. However, there are some areas where CSLs have been successfully exploited in a grain boundary engineering context. These are:

- One- or two-dimensional polycrystals (e.g. wires or sheets) having a strong texture, which allow CSLs with a specific misorientation axis and a favourable boundary plane to develop;
- Materials containing a high proportion of low angle boundaries ($\Sigma 1$);
- Materials containing a high proportion of twins ($\Sigma 3$).

There are circumstances where the CSL alone is a necessary but not sufficient predictor of special properties. For example:

- Knowledge of the Σ-value alone does not predict the location of the boundary plane – a key element in boundary geometry and properties;
- The CSL is a *geometric model* and does not account for either the atomic level or the extrinsic structure of the boundary, which has an effect on properties;
- Factors other than the intrinsic or extrinsic structure of the boundary may affect its behaviour, e.g. the orientation of the interfacing grains, the orientation of external features such as a tensile axis, or particularly extreme environmental conditions.

Finally, what are the areas in which progress can be made with respect to the CSL model and grain boundary engineering? The *range* of materials to which the CSL is applied, in a modified form, is extending to other classes such as intermetallics and superconductors. The importance of *clustering* and *connectivity* of boundaries in microstructures is now realised and is being studied by computer simulation and in real microstructures. And probably the most important element for continued progress in grain boundary engineering is to be able to predict and manipulate the orientation of the *grain boundary plane* within the CSL. More in-depth information of these types, coupled with realistic expectations of what the CSL can deliver, will augment its application to improvement of grain boundary performance.

References

1. M.L. Kronberg and F.H. Wilson, *Met. Trans.* **185** (1949) 501.
2. K.T. Aust and J.W. Rutter, *Trans. TMS-AIME* **215** (1959) 119.
3. G. Herrmann, H. Gleiter and G. Baro, *Acta Metall.* **24** (1976) 353.
4. A.P. Sutton and R.W. Baluffi, *Acta Metall.* **35** (1987) 2177.
5. L Priester, *Revue Phys. Appl.* **24** (1989) 419.
6. S.C. Mehta and D.A. Smith, Proc. *Grain boundary engineering*, Ed. U. Erb and G. Palumbo, Canadian. Inst. Min., Met. & Petrol., Montreal (1994) 89.
7. P.J. Goodhew and D.A. Smith, *Scripta Met.* **14** (1980) 59.
8. R.A. Varin, *Phys. Stat. Sol.* **51A** (1979) k189.
9. L.S. Shvindlerman and B.B. Straumal, *Acta Metall.* **33** (1985) 1735.
10. T. Aust, U. Erb and G. Palumbo, in *Mechanical properties and deformation behaviour of materials having ultra-fine microstructures*, Ed. M. Nastasi et al, Kluwer Academic Publishers (1993) 107.
11. T. Watanabe, *Res. Mech.* **11** (1984) 47.
12. P. Lin, G. Palumbo, U. Erb and K.T. Aust, *Scripta Met.* **33** (1995) 1387.
13. V. Randle, *Microtexture determination and its applications*, Institute of Materials, London, (1992).
14. R.A. Schwarzer, *Tex. and Micros.* **20** (1993) 7.
15. A. Berger, P.J. Wilbrandt and P. Haasen, *Acta Metall.* **31** (1983) 1443.
16. D.J. Dingley, *Mat. Sci. Tech.* in press.
17. V. Randle, *The measurement of grain boundary geometry*, Inst. Physics Pub., Bristol, UK (1993).
18. C. Goux, *Can. Metall. Q.* **13** (1974) 9.
19. D. Wolf, *Phil. Mag.* **A62** (1990) 447.
20. F.F. Lange, *Acta Metall.* **15** (1967) 311.
21. V. Randle, *Acta Metall. Mat.* **42** (1994) 1769.
22. G.H. Bishop and B. Chalmers, *Scripta Met.* **2** (1968) 133.
23. P.H. Pumphrey and K.M. Bowkett, *Scripta Met.* **6** (1972) 31.
24. H. Grimmer, W. Bollmann and D.H. Warrington, *Acta Cryst.* **A30** (1974) 197.
25. G. Friedel, *Leçons de cristallographie*, Ed. Blanchard (1926).
26. S. Ranganathan, *Acta Cryst.* **21** (1966) 197.
27. D.G. Brandon, B. Ralph, S. Ranganathan and M.S. Wald, *Acta Metall.* **12** (1964) 813.
28. W. Bollmann, *Crystalline defects and crystalline interfaces*, Springer, Berlin (1970).
29. J. Levy, *Phys. Stat. Sol.* **31** (1969) 193.
30. Y. Ishida, T. Hasegawa and F. Nagata, *J. Appl. Cryst.* **40** (1969) 2182.
31. C.T. Forwood and L.M. Clarebrough, *Electron microscopy of interfaces in metals and alloys*, Inst. Physics Pub, Bristol, UK (1992).
32. B. Chalmers and H. Gleiter, *Phil. Mag.* **23** (1971) 1541.
33. M.F. Ashby, F. Spaepen and S. Williams, *Acta Metall.* **26** (1978) 1674.
34. R.C. Pond, V. Vitek and D.A. Smith, *Acta Cryst.* **A35** (1979) 689.
35. A.P. Sutton and V. Vitek, *Philos. Trans. R. Soc. Lond.* **309A** (1983) 1.

36. R.W. Balluffi and P.D. Bristowe, *Surf. Sci.* **144** (1984) 28.
37. H. Mykura, in *Grain boundary structure and kinetics*, ASM, Ohio, USA (1980) 445.
38. A. Garg, W.A.T. Clark and J.P. Hirth, *Phil. Mag.* **A59** (1989) 479.
39. D.H. Warrington and M. Boon, *Acta Met.* **23** (1975) 599.
40. S. Chan, J. *Phys. Chem. Sol.* **55** (1994) 1137.
41. W. Chen and F. Chen, *Ultramicroscopy* **51** (1993) 316.
42. Y. Zhu, H. Zhang, H. Wang and M. Suenaga, *J. Mat. Res.* **6** (1991) 2507.
43. R. Bonnet and F. Durand, *Phil. Mag.* **32** (1975) 997.
44. W.T. Read and W. Shockley, *Phys. Rev.* **78** (1950) 275.
45. P. Pumphrey, PhD thesis, Cambridge (1974).
46. D.G. Brandon, *Acta Metall.* **14** (1966) 1479.
47. G. Palumbo and K.T. Aust, *Materials interfaces: atomic level structure and properties.* Ed. D. Wolf and S. Yip, Chapman and Hall, (1992) 190.
48. D. Wolf, *J. de Phys.* **46** (1985) C4–197.
49. D. Wolf, *Acta Metall. Mat.* **38** (1990) 791.
50. K.L. Merkle and D. Wolf, *Phil. Mag.* **65A** (1992) 513.
51. K.L. Merkle and D. Wolf, *Mat. Res. Soc. Symp. Proc.,* **229**, The Materials Research Soc., (1991) 185.
52. G.C. Hasson and C. Goux, *Scripta Met.* **5** (1971) 889.
53. P. Chaudari and J.W. Matthews, *J. Appl. Phys.* **42** (1971) 3063.
54. I. Nakamichi, *J. Sci. Hiroshima Univ.,* **54A** (1990) 49.
55. A.P. Sutton and V. Vitek, *Phil. Trans. R. Soc. Lond.* **A309** (1983) 55.
56. G. Gottstein and L.S. Shvindlerman, *Scripta Met. Mat.* **27** (1992) 1515.
57. G. Gottstein and L.S. Shvindlerman, *Scripta Met. Mat.* **27** (1992) 1521.
58. R.A. Varin and K. Tangri, *Met. Trans.* **12A** (1981) 1859.
59. A. Garbacz and J.W. Wyrzykowski, *Mat. Sci. Eng.* **A172** (1993) 137.
60. D. Mclean, *Phil. Mag.* **23** (1971) 467.
61. S. Lartigue and L. Priester, *Acta Metall.* **31** (1983) 1809.
62. V. Randle and A. Brown, *Phil. Mag.* **59A** (1989) 1075.
63. P. Gordon and R.A. Vandermeer, in *Recrystallisation, grain growth and textures*, Ed. H. Margolin, ASM Ohio (1965) 205.
64. V. Randle and B. Ralph, *Rev. Phys. Appl.* **23** (1988) 501.
65. *Materials interfaces: atomic level structure and properties*, Ed. D. Wolf and S. Yip, Chapman and Hall, London, (1992).
66. L.K. Fionova, *Mat. Chem. & Phys.,* **37** (1994) 201.
67. R.W. Balluffi and R. Maurer, *Scripta Met.* **22** (1988) 709.
68. H. Mykura, *Acta Metall.* **27** (1979) 243.
69. G. Gunther, P.J. Wilbrandt and P. Haasen, *Mat. Sci. For.* **113–115** (1993) 715.
70. W. Lojowski, H. Gleiter and R. Maurer, *Acta Metall.* **36** (1988) 69.
71. M. Dechamps, F. Barbier and A. Marrouche, *Acta Metall.* **35** (1987) 101.
72. C.S. Nichols, R.F. Cook, D.R. Clarke and D.A. Smith, *Acta Met. Mat.* **39** (1991) 1657 and 1667.
73. V. Randle, Int. Symp. *Grain boundary engineering*, Ed. U. Erb and G. Palumbo, Can. Inst. Min. Met. & Petrol, Montreal, (1995) 89.
74. V. Randle, Proc. ICOTOM10, Ed. H.J. Bunge, *Mat. Sci. For.* **157–162** (1995) 175.
75. R. Omar, PhD thesis, University of Warwick, UK (1987).

76. W. Jaeger and H. Gleiter, *Scripta Met.* **12** (1978) 675.

77. H. Lin and D.P. Pope, *Acta Met. Mat.* **41** (1993) 553.

78. B. Dorner, P.J. Wilbrandt and P. Haasen, Proc. ICOTOM10, Ed. H.J. Bunge, *Mat. Sci. For.* **157–162** (1995) 175.

79. L.S. Shvindlerman, V.G. Sursaeva, V.P. Yashnikov and R.G. Faulkner, Proc. ICOTOM10, Ed. H.J. Bunge, *Mat. Sci. For.* **157–162** (1995) 175.

80. T. Watanabe, H. Fujii, H. Oikawa and K.I. Arai, *Acta Metall.* **37** (1989) 941.

81. H.J. Bunge and H. Weiland, *Tex. & Micros.* **7** (1988) 231.

82. J. Zhao, B.L. Adams and P.R. Morris, *Tex. & Micros.* **8/9** (1988) 493.

83. B.L. Adams, *Met. Trans.* **17A** (1986) 2199.

84. B.L. Adams, J. Zhao and D. O'Hara, *Acta Met. Mat.* **38** (1990) 953.

85. Y. Pan and B.L. Adams, *Scripta Met. Mat.* **30** (1994) 1055.

86. A. Garbacz and M.W.Grabski, *Scripta Met.* **23** (1989) 1369.

87. V.Y. Gertsman, A.P. Zhilyaev, A.I. Pshenichnyuk and R.Z. Valiev, *Acta Met. Mat.* **40** (1992) 1433.

88. V. Randle, *Mat. Sci. Tech.* **6** (1990) 1231.

89. V. Randle, *Mod. Sim. Mat. Sci. Eng.*, in press.

90. P. Fortier, K.T. Aust and W.A. Miller, *Acta Met. Mat.* **43** (1995) 339.

91. V. Randle, *Scripta Met. Mat.* **28** (1993) 889.

92. V. Randle, *Acta Met. Mat.* **43** (1995) 1741.

93. T. Watanabe, *Mat. Sci. Eng.* **A176** (1994) 39.

94. K.T. Lee and J.A. Szpunar, Proc. ICOTOM10, Ed. H.J. Bunge, *Mat. Sci. For.* **157–162** (1995) 989.

95. H.M. Kim and J.A. Szpunar, Proc. IOCTOM10, Ed. H.J. Bunge, *Mat. Sci. For.* **157–162** (1995) 1997.

96. V. Randle, *Tex. & Micros.* **20** (1993) 231.

97. B.L. Adams, D.J. Dingley, K. Kunze and S.I. Wright, *Mat. Sci. For.* **157–162** (1994) 31.

98. T. Watanabe, T. Hirano, T. Ochiai and H. Oikawa, Proc. IOCTOM10, Ed. H.J. Bunge, *Mat. Sci. For.* **157–162** (1995) 1103.

99. R. Shimizu, J. Harase and D.J. Dingley, *Acta Met. Mat.* **38** (1990) 973.

100. T. Watanabe, *Scripta Met.* **27** (1992) 1497.

101. K.T. Aust, in *'Grain boundary engineering'* Ed. U. Erb and G. Palumbo, Can. Inst. Min. Met. & Petrol., Montreal, Canada (1994) 197.

102. F. Haessner, K. Sztwiertnia and P.J. Wilbrandt, *Tex. & Micros.* **13** (1991) 213.

103. V. Randle, *Proc. Roy. Soc. Lond.* **431A** (1990) 61.

104. V. Randle and A. Day, *Mat. Sci. Tech.* **9** (1993) 1069.

105. J. Zhao, J.S. Koontz and B.L. Adams, *Met. Trans.* **19A** (1988) 1179.

106. D.P. Field, Proc. IOCTOM10, Ed. H.J. Bunge, *Mat. Sci. For.* **157–162** (1995).

107. J. Zhao and B.L. Adams, *Acta Cryst.* A44 (1988) 326.

108. D.P. Field and B.L. Adams, *Acta Met. Mat.* **40** (1992) 1145.

109. A. Garbacz and M.W. Grabski, *Acta Met. Mat.* **41** (1993) 469, 475.

110. V. Randle and B. Ralph, *Tex. & Micros.* **8/9** (1988) 531.

111. A. Morawiec, J.A. Szpunar and D.C. Hinz, *Acta Met. Mat.* **41** (1994) 2825.

112. V. Randle, Proc. ICOTOM11, in press.

113. V.Y. Gertsman and K. Tangri, *Acta Met. Mat.* **43** (1995) 2317.

114. P. Haasen, *Scripta Met. Mat.* **27** (1992) 1477.

115. M. Heinrich, P.J. Wilbrandt and P. Haasen, Proc. IOCTOM10, Ed. H.J. Bunge, *Mat. Sci. For.* **157–162** (1995) 965.

116. G. Palumbo and K.T. Aust, Proc. *Recrystallisation '90*, Ed. T. Chandra, Min. Met. Mat. Soc., (1990) 101.

117. T. Ogura, T. Watanabe, S. Karashima and T. Masumoto, *Acta Metall.* **35** (1987) 1807.

118. D. Farkas, M.O. Lewus and V. Rangarajan, *Scripta Met.* **22** (1988) 1195.

119. S.R. Ortner and V. Randle, *Scripta Met.* **23** (1989) 1903.

120. L.C. Lim and R. Raj, *Acta Metall.* **32** (1984) 1183.

121. H. Lin and D.P. Pope, *Acta Met. Mat.* **41** (1993) 553.

122. V. Randle, *Tex. & Micros.* **20** (1993) 231.

123. L.E. Murr, *J. Appl. Phys.* **39** (1968) 5557.

124. V. Randle, *Micros., Microanal., Microstruct.* **4** (1993) 349.

125. J. Harase and R. Shimizu, *7th Risø Int. Symp. on Annealing*, Ed. N. Hansen et al, Risø press, Denmark, (1986) 343.

126. C.B. Thomson and V. Randle, *J. Mat. Sci.*, in press.

127. P. Gangli and J. Szpunar, Proc. ICOTOM10, Ed. H.J. Bunge, *Mat. Sci. For.* **157–162** (1994) 927.

128. T. Urabe and J.J. Jonas, *ISIJ International* **34** (1994) 435.

129. K.L. Merkle, *J. Phys. Chem. Sol.* **55** (1994) 991.

130. V. Randle, *Proc. 52nd meeting MSA*, Ed. G.W. Bailey and A.J. Garratt-Read, San Francisco Press Inc., (1994) 616.

131. V. Paidar, *Phil. Mag.* **A66** (1992) 41.

132. D. Bouchet and L. Priester, *Scripta Met.* **21** (1987) 475.

133. K.L. Merkle, *Ultramicroscopy* **37** (1991) 130.

134. U. Wolf, F. Ernst, T. Muschik, M.W. Finnis and H.F. Fischmeister, *Phil. Mag.* **66A** (1992) 991.

135. R.W. Balluffi and R. Maurer, *Scripta Met.* **22** (1988) 709.

136. V. Randle, *Mat. Charact.* **34** (1995) 29.

137. V. Randle, *Scripta Met.* **23** (1989) 773.

138. V. Randle and D.J. Dingley, Proc. Euromat '89, *Advanced materials and processes*, Ed. H.E. Exner and V. Schumacher, DGM Oberursel, Germany (1990) 1273.

139. V. Randle, *Mat. Sci. Tech.* **7** (1991) 985.

140. V. Randle, *J. Mat. Sci.* **30** (1995) 3983.

141. V. Randle, *Acta Cryst.* **A50** (1994) 588.

142. A. Garbacz, B. Ralph and K.J. Kurzydlowski, *Acta Met. Mat.* **43** (1995) 1541.

143. G. Palumbo, K.T. Aust, U. Erb, P.J. King, A.M. Brennenstuhl and P.C. Lichtenberger, *Phys. Stat. Sol.* **131** (1992) 425.

144. V.Y. Gertsman and K. Tangri, *Phil. Mag.* **A64** (1991) 1319.

145. J. Furley and V. Randle, *Mat. Sci. Tech.* **7** (1991) 12.

146. C.T. Forwood and L.M. Clarebrough, *Acta Metall.* **32** (1984) 757.

147. V.Y. Gertsman and K. Tangri, *Scripta Met.* **32** (1995) 1649.

148. C.V. Kopezky, A.V. Andreeva and G.D. Sukhomlin, *Acta Met. Mat.* **39** (1991) 1603.

149. V. Randle, *Proc. Conf. on X-rays and related techniques*, Malaysia, (1996) University Sains Malaysia, 11.

150. M.C. Carl, J. Fiedler and V. Randle, *Scripta Met. Mat.*, in press.

151. P. Haasen and P.J. Wilbrandt, Proc. IOCTOM10, Ed. H.J. Bunge, *Mat. Sci. For.* **157–162** (1995) 887.

152. A. Berger, P.J. Wilbrandt, F. Ernst, U. Klement and P. Haasen, Prog. *Mat. Sci.* **32** (1988) 1.

153. C. Kunke, P. Haasen, U. Klement and P.J. Wilbrandt, *Mat. Sci. For.* **113–115** (1993) 133.

154. K. Sztwiertnia and F. Haessner, Proc. IOCTOM10, Ed. H.J. Bunge, *Mat. Sci. For.* **157–162** (1995) 1069.

155. F. Haessner, K. Sztwiertnia and P.J. Wilbrandt, *Tex. & Micros.* **13** (1991) 213.

156. V. Randle, D. Juul Jensen and N. Hansen, *Phil. Mag.*, **73** (1996) 265.

157. R. Shimizu and J. Harase, *Acta Metall.* **37** (1989) 1241.

158. J. Harase, Y. Ushigami and N. Takahashi, *Mat. Sci. For.* **157–162** (1994) 899.

159. R. Shimizu, J. Harase and D.J. Dingley, *Acta Met. Mat.* **38** (1990) 973.

160. J. Harase, *Mat. Sci. For.* **94–96** (1992) 419.

161. V. Randle and J. Furley, *Tex. & Micros.* **14/18** (1991) 877.

162. V.Y. Gertsman and K. Tangri, *Acta Met. Mat.* **43** (1995) 2317.

163. R.L. Fullman and J.C. Fisher, *J. Appl. Phys.* **22** (1951) 1350.

164. M. Meyers and L. Murr, *Acta Metall.* **26** (1978) 951.

165. D. Horton, C.B. Thomson and V. Randle, *Mat. Sci. Eng.* **A203** (1995) 408.

166. S.G. Lewis and V. Randle, *Tex. & Micros.*, in press.

167. V. Randle and A. Brown, *Phil. Mag.* **A58** (1988) 717.

168. C.B. Thomson and V. Randle, *Tex. & Micros.*, in press.

Appendix

Alphabetical list of the investigations which have been included to compile the CSL statistics

B.L. Adams, J.W. Zhao and D. O'Hara, *Acta Met. Mat.* **38** (1990) 953.

M.H. Ainsley, G.J. Cocks and D.R. Miller, *Met. Sci.* **13** (1979) 20.

A.V. Andrejeva, G.I. Salnikov and L.K. Fionova, *Acta Metall.* **26** (1978) 1331.

K.T. Aust, U. Erb and G. Palumbo, *Mat. Sci. Eng.* **A176** (1994) 329.

K.T. Aust, *Can. Metall. Quart.* **33** (1994) 265.

K.T. Aust, U. Erb and G. Palumbo, *Mechanical properties and deformation behaviour of materials having ultra-fine microstructures*, Ed. M. Nastasi et al, Kluwer Academic Publishers, Holland, (1993) 107.

K.T. Aust, *Int. Symp. on Grain boundary engineering*, Ed. U. Erb and G. Palumbo, Canadian Inst. Min. Met. & Petrol., Montreal, (1994) 197.

G. Avramovic-Cingara, K.T. Aust, D.D. Perovic, G. Palumbo and H.J. McQueen, *Proc. Inst. Symp. on Light Metals*, Ed. M. Gilbert et al, Met. Soc. CIM, (1994) 281.

J. Ball and G. Gottstein, *Mat. Sci. For.* **113–115** (1993) 569.

P.T. Baudin, P. Paillard and R. Penelle, *J. Appl. Cryst.* **25** (1992) 400.

R.V. Belluz and K.T. Aust, *Met. Trans.* **6A** (1975) 219.

A. Berger, P.J. Wilbrandt, F. Ernst, U. Klement and P. Haasen, *Prog. Mat. Sci.* **32** (1988) 1.

A. Berger, P.J. Wilbrandt and P. Haasen, *Acta Metall.* **9** (1983). 1433.

D. Bouchet and L. Priester, *Scripta Met.* **21** (1987) 475.

D. Bouchet and L. Priester, *Scripta Met.* **20** (1986) 961.

S. Brodesser, G. Bruckner and G. Gottstein, Proc. ICOTOM10, Ed. H.J. Bunge, *Mat. Sci. For.* **157–162** (1994) 1153.

S. Brodesser, S. Chen and G. Gottstein, Proc. ICOTOM9, *Tex. & Micros.* **14/18** (1991) 829.

S. Brodesser and G. Gottstein, Proc. Conf. *Microscale textures of materials*, Ed. B.L. Adams and H.J. Bunge, *Tex. & Micros.* **20** (1993) 179.

M.P. Butron-Guillen, J.G. Cabanas-Moreno and J.R. Weeterman, *Scripta Met. Mat.* **24** (1990) 991.

D.C Crawford and G.S. Was, *Met. Trans.* **23A** (1992) 1195.

D.C. Crawford and G.S. Was, *J. Elect. Mic. Tech.* **19** (1991) 345.

P. Czurratis, R. Kroggel and H. Loffler, *Zeit. Metall.* **79** (1988) 307.

M. Dechamps, Proc. ICOTOM9, *Tex. & Micros.* **14–18** (1991) 733.

M. Dechamps, A. Marrouche, F. Barbier and A. Revcolevschi, *J. de Phys.* **46** (1985) C4–435.

M. Dechamps, Proc. ICOTOM9, *Tex. & Micros.* **14/18** (1991) 733.

J. Don and S. Majumdar, *Acta Metall.* **34** (1986) 961.

B. Dorner, P.J. Wilbrandt and P. Haasen, Proc. ICOTOM10, Ed. H.J. Bunge, *Mat. Sci. For.* **157–162** (1994) 927.

B. El M'Rabat and L. Priester, *Mat. Sci. Eng.* **101A** (1988) 117.

D. Farkas, M.O. Lewus and V. Rangarajan, *Scripta Met.* **22** (1988) 1195.

112   *Appendix*

D.P. Field and B.L. Adams, *Proc. Conf. Microscale textures of materials*, Ed. B.L. Adams and H.J. Bunge, *Tex. & Micros.* **20** (1993) 217.

D.P. Field, Proc. ICOTOM10, Ed. H.J. Bunge, *Mat. Sci. For.* **157–162** (1994) 1175.

D.P. Field and B.L. Adams, *Acta Met. Mat.* **40** (1992) 1145.

L.K. Fionova, J.A. Juarez-Islas, R. Perez, J.L. Albarran, O. Flores, L. Martinez and Y. Titovets, *Scripta Met. Mat.* **31** (1994) 1619.

L.K. Fionova, O. Konokenko, V. Matveev, L. Priester, S. Lartigue and F. Dupau, *Inter. Sci.* **1** (1993) 207.

L.K. Fionova, *Mat. Chem. Phys.,* **37** (1994) 201.

C. Funke, P. Haasen, U. Klement and P.J. Wilbrandt, *Mat. Sci. For.* **113–115** (1993) 133.

A.W. Funkenbusch and A.F. Giamei, *Conf. on Interface migration and control of microstructure*, Ed. C. Pande et al, ASM, Ohio, USA (1986) 161.

J. Furley and V. Randle, *Mat. Sci. Tech.* **7** (1991) 12.

P. Gangli, J.A. Szpunar and F. Golini, Proc. ICOTOM9, *Tex. & Micros.* **14/18** (1991) 819.

P. Gangli and J.A. Szpunar, Proc. ICOTOM10, Ed. H.J. Bunge, *Mat. Sci. For.* **157–162** (1994) 953.

A. Garbacz and J.W. Wyrzykowski, *Mat. Sci. Eng.* **A172** (1993) 137.

A. Garbacz and M.W. Grabski, *Acta Met. Mat.,* **41** (1993) 475.

A. Garbacz, B. Ralph and K.J. Kurzydlowski, *Acta Met. Mat.* **43** (1995) 1541.

A. Garbacz and M.W. Grabski, *Acta Met. Mat.* **41** (1993) 469.

A. Garbacz and M.W. Grabski, *Scripta Met.* **23** (1989) 1369.

A. Garg, W.A.T. Clark and J.P. Hirth, *Phil. Mag.* **A59** (1989) 479.

D. Gerth and R.A. Schwarzer, *Mat. Sci. For.* **113–115** (1993) 619, 625.

V.Y. Gertsman and K. Tangri, *Phil. Mag.* **A64** (1991) 1319.

V.Y. Gertsman and K. Tangri, *Acta Met. Mat.* **43** (1995) 2317.

V.Y. Gertsman and K. Tangri, *Scripta Met. Mat.* **33** (1995) 1037.

V.Y. Gertsman, A.P. Zhilyaev, A.I. Pshenichnyuk and R.Z. Valiev, *Acta Met. Mat.* **40** (1992) 1433.

G. Gunther, P.J. Wilbrandt and P. Haasen, *Mat. Sci. For.* **113–115** (1993) 661.

P. Haasen, Proc. *7th Risø Int. Symp., Annealing*, Ed. N. Hansen et al, Risø Press, Denmark (1986) 69.

P. Haasen, *Scripta Met. Mat.* **27** (1992) 1477.

P. Haasen, Proc. *Recrystallisation '90*, Ed. T. Chandra, Met. Soc. AIME, Warrendale, USA (1990) 17.

P. Haasen and P. J. Wilbrandt, Proc. ICOTOM10, Ed. H.J. Bunge, *Mat. Sci. For.* **157–162** (1994) 887.

F. Haessner, K. Sztwiertnia and P.J. Wilbrandt, *Tex. & Micros.* **13** (1991) 213.

S. Hanada, T. Ogura, S. Watanabe, O. Izumi and T. Masumoto, *Acta Metall.* **34** (1986) 13.

J. Harase and R. Shimizu, *7th Risø Int. Symp., Annealing*, Ed. N. Hansen et al, Risø Press, Denmark (1986) 343.

J. Harase, R. Shimizu and T. Watanabe, *10th Risø Int. Symp., Materials architecture*, Ed. J.B. Bilde-Sorensen et al, Risø Press, Denmark (1989) 365.

J. Harase, Y. Ushigami and N. Takahashi, *Mat. Sci. For.* **157–162** (1994) 899.

J. Harase and R. Shimizu, *Acta Met. Mat.* **38** (1990) 1395.

J. Harase, R. Shimizu, Y. Nakamura and N. Takahashi, *Mat. For.* **14** (1990) 276.

J. Harase, R. Shimizu and N. Takahashi, Proc. ICOTOM9, *Tex. & Micros.* **14/18** (1991) 679.

J. Harase, Proc. Conf. *Grain growth in polycrystalline materials*, Mat. Sci. For. **94−96** (1992) 419.

J. Harase and R. Shimizu, Proc. Conf. *Grain growth in polycrystalline materials*, Mat. Sci. For. **94−96** (1992) 399.

J. Harase, *Mat. Sci. For.* **94−96** (1992) 419.

J. Harase, K. Tomimasu, S. Nakashima, T. Nagashima, K. Takashima, Y. Yoshitomi and N. Takahashi, *Mat. Sci. For.* **113−115** (1993) 709.

J. Harase, Y. Ushigami and N. Takahashi, Proc. ICOTOM10, Ed. H.J. Bunge, *Mat. Sci. For.* **157−162** (1994) 899.

C. Heidelbach, J. Pospiech and H.R. Wenk, Proc. ICOTOM10, Ed. H.J. Bunge, *Mat. Sci. For.* **157−162** (1994) 965.

M. Heinrich, P.J. Wilbrandt and P. Haasen, Proc. ICOTOM10, Ed. H.J. Bunge, *Mat. Sci. For.* **157−162** (1994) 971.

H.W.F. Heller, C.A. Verbraak and B.H. Kolster, *Acta Metall.* **32** (1984) 1395.

D. Horton, C.B. Thomson and V. Randle, *Mat. Sci. Eng.* **A203** (1995) 408.

W. Jaeger and H. Gleiter, *Scripta Met.* **12** (1978) 675.

W.A. Kaysser, S. Takajo and G. Petzow, *Bd.* **73** (1982) 579.

H.M. Kim and J.A. Szpunar, Proc. ICOTOM10, Ed. H.J. Bunge, *Mat. Sci. For.* **157−162** (1994) 1997.

U. Klement and P. Haasen, *Acta Met. Mat.* **41** (1993) 1075.

C.V. Kopezky and L.K. Fionova, *Proc. Recrystallisation '90*, Ed. T. Chandra, Met. Soc. AIME, Warrendale, USA (1990) 255.

C.V. Kopezky, A.V. Andreeva and G.D. Sukhomlin, *Acta Met. Mat.* **39** (1991) 1603.

M.S. Laws and P.J. Goodhew, *Acta Met. Mat.* **39** (1991) 1525.

W.B. Lee, J. Furley and B. Ralph, *J. Mat. Sci.* **27** (1992) 3395.

K.T. Lee and J.A. Szpunar, Proc. ICOTOM10, Ed. H.J. Bunge, *Mat. Sci. For.* **157−162** (1994) 989.

K.T. Lee, G. deWitt, A. Morawiec and J.A. Szpunar, *J. Mat. Sci.* **30** (1995) 1327.

L.C. Lim, *Acta Metall.* **35** (1987) 1653.

L.C. Lim. and R. Raj, *Acta Metall.* **32** (1984) 1183.

L.C. Lim and R. Raj, *Acta Metall.* **32** (1984) 1177.

H. Lin and D.P. Pope, *Acta Met. Mat.* **41** (1993) 553.

P. Lin, G. Palumbo, U. Erb and K.T. Aust, *Scripta Met. Mat.* **33** (1995) 1387.

W. Liu, M. Bayerlein, H. Mughrabi, A. Day and P.N. Quested, *Acta Met. Mat.* **40** (1992) 1763.

H. Liu, M. Gao, D.G. Harlow and R.P. Wei, *Scripta Met. Mat.* **32** (1995) 1807.

R.A.D. Mackenzie, M.D. Vaudin and S.L. Sass, in *Proc. 46th meeting EMSA*, Ed. G.W. Bailey, San Francisco Press Inc. (1988) 602.

H. Makita, S. Hanada, O. Izumi, H. Fukuda and T. Imaizumi, *Proc. Recrystallisation '90*, Ed. T. Chandra, Met. Soc. AIME, Warrendale, USA (1990) 617.

H. Makita, S. Hanada and O. Izumi, *Acta Met.* **36** (1988) 403.

H.O. Martikainen and V.K. Lindroos, *Acta Metall.* **33** (1995) 1223.

T.A. Mason and B.L. Adams, *Proc. Conf. Microscale textures of materials*, Ed. B.L. Adams and H.J. Bunge, *Tex. & Micros.* **20** (1993) 111.

K. Matsumoto, T. Shibayanagi and Y. Umakoshi, *Scripta Met. Mat.* **33** (1995) 1321.

D.E. Meyers and A.J. Ardell, *Acta Met. Mat.* **41** (1993) 2601.

J. Mizera, A. Garbacz and K.J. Kurzydlowski, *Scripta Met. Mat.* **33** (1995) 515.

A. Morawiec, J.A. Szpunar and D.C. Hinz, *Acta Met. Mat.* **41** (1993) 2825.

A. Morawiec and J.A. Szpunar, Proc. ICOTOM10, Ed. H.J. Bunge, *Mat. Sci. For.* **157–162** (1994) 1263.

T. Ogura, T. Watanabe, S. Karashima and T. Masumoto, *Acta Metall.* **35** (1987) 1807.

R. Omar, PhD Thesis, University of Warwick, UK (1987).

R. Omar and H. Mykura, in *Proc. Symp. Interfacial structure, properties and design,* **122,** Ed. M.H. Yoo et al, MRS, Pittsburgh, USA (1988).

S. Ortner and V. Randle, *Scripta Met.* **23** (1989) 1903.

G. Palumbo and K.T. Aust, in *Materials interfaces: atomic level structure and properties,* Ed. D. Wolf and S. Yip, Chapman & Hall, London, (1992) 190.

G. Palumbo and K.T. Aust, *Proc. Recrystallisation '90,* Ed. T. Chandra, Met. Soc. AIME, Warrendale, USA (1990) 101.

G. Palumbo and K.T. Aust, *Acta Met. Mat.* **38** (1990) 2343.

G. Palumbo and K.T. Aust, *Scripta Met.,* **22** (1988) 847.

Y. Pan and B.L. Adams, *Scripta Met. Mat.* **30** (1994) 1055.

R. Penelle, T. Baudin, P. Paillard and L. Mora, Proc. ICOTOM9 *Tex. & Micros.* **14–18** (1991) 597.

D. Raabe and K. Lucke, *Scripta Met. Mat.* **26** (1992) 19.

L. Rabet, L. Kestens, P. Van Houtte and E. Aernoudt, *Proc. Grain growth in polycrystalline materials, Mat. Sci. For.* **94–96** (1992) 611.

V. Raman, T. Watanabe and T.G. Langdon, *Acta Metall.* **37** (1989) 705.

V. Randle and B. Ralph, *Tex. & Micros.* **8/9** (1988c) 531.

V. Randle and D.J. Dingley, Proc. *Euromat '89,* Ed. H.E. Exner and V. Schumacher, DGM Oberursel, Germay (1990) 1273.

V. Randle, *Scripta Met.* **23** (1989a) 773.

V. Randle, Proc. *10th Risø Int. Symp.,* Ed. J.B. Bilde-Sorensen et al, Risø Press, Denmark (1989b) 529.

V. Randle, *Acta Met. Mat.* **39** (1991a) 481.

V. Randle, *Met. Trans.* **21A** (1990b) 2215.

V. Randle and B. Ralph, *Rev. Phys. Appl.* **23** (1988a) 501.

V. Randle and B. Ralph, *Proc. EMAG '87,* Inst. Phys. Conf. Ser. no. 90 (1987b) 205.

V. Randle, *Mat. Sci. Tech.* **7** (1991b) 985.

V. Randle, *Phil. Mag.* **67A** (1993) 1301.

V. Randle, *Acta Met. Mat.* **43** (1995a) 1741.

V. Randle, *Acta Cryst.* **A50** (1994) 588.

V. Randle and A. Brown, *Phil. Mag.* **A58** (1988) 717 and **A59** (1989) 1075.

V. Randle, *J. Mat. Sci.* **30** (1995b) 3983.

V. Randle and B. Ralph, *J. Mat. Sci.* **22** (1987a) 2535.

V. Randle and S.G. Lewis, *Tex. & Micros.,* in press.

V. Randle and B. Ralph, *Proc. Roy. Soc. Lond.,* **A415** (1988b) 239.

V. Randle, B. Ralph and D.J. Dingley, *Acta Metall.* **36** (1988) 267.

V. Randle, *Mod. Sim. Mat. Sci. Eng.* in press.

V. Randle, *Mat. Sci. Tech.* **6** (1990a) 1231.

V. Randle and D.J. Dingley, *Scripta Met.* **23** (1989) 1565.

V. Randle, *Int. Symp. on Grain boundary engineering*, Ed. U. Erb and G. Palumbo, Canadian Inst. Min. Met. & Petrol., Montreal, (1994) 89.

N. Rouag, G. Vigna and R. Penelle, *Acta Met. Mat.* **38** (1990) 1101.

N. Rouag and P. Penelle, *Tex. & Micros.* **11** (1989) 203.

A. Sato, K. Kon, S. Tsujikawa and Y. Hisamatsu, in *Proc. JIMIS-4, Grain boundary structure and related phenomena*, Supp. to Trans. JIM, (1986) 773.

T. Shibayanagi, H. Takatani and S. Hori, *Proc. Conf. Grain growth in polycrystalline materials, Mat. Sci. For.* **94–96** (1992) 495.

R. Shimizu, J. Harase and D.J. Dingley, *Acta Met. Mat.* **38** (1990) 973.

R. Shimizu and J. Harase, *Acta Metall.* **37** (1989) 1241.

L.S. Shvindlerman, V.G. Sursaeva, V.P. Yashnikov and R.G. Faulkner, Proc. ICOTOM10, Ed. H.J. Bunge, *Mat. Sci. For.* **157–162** (1994) 1057.

K. Sztwiertnia and F. Haessner, *Proc. Recrystallisation '90*, Ed. T. Chandra, Met. Soc. AIME, Warrendale, USA (1990) 535.

K. Sztwiertnia and F. Haessner, Proc. ICOTOM9, *Tex. & Micros.* **14/18** (1991) 641.

K. Sztwiertnia and F. Haessner, Proc. ICOTOM10, Ed. H.J. Bunge, *Mat. Sci. For.* **157–162** (1994) 1069.

C.B. Thomson and V. Randle, *J. Mat. Sci.*, in press.

C.B. Thomson and V. Randle, *Tex. & Micros.*, in press.

C.J. Tweed, B. Ralph and N. Hansen, *Acta Metall.* **32** (1984) 1407.

P.N.T. Unwin and R.B. Nicholson, *Acta Met.* **17** (1969) 1379.

T. Urabe and J.J. Jonas, *ISIJ Int.* **34** (1994) 435.

Y. Ushigami, K. Kawasaki, T. Nakayama, Y. Suga, J. Harase and N. Takahashi, Proc. ICOTOM10, Ed. H.J. Bunge, *Mat. Sci. For.* **157–162** (1994) 1081.

R.A. Varin, *Phys. Stat. Sol.,* **A51** (1979) K189.

R.A. Varin, J.W. Wyrzykowski, W. Lojkowski and M.W. Grabski, *Phys. Stat. Sol.* **A45** (1978) 565.

R.A. Varin and K. Tangri, *Met. Trans.* **12A** (1981) 1859.

D. Vaughan, *Acta Metall.* **18** (1970) 183.

T. Watanabe, T. Hirano, T. Ochiai and H. Oikawa, Proc. ICOTOM10, Ed. H.J. Bunge, *Mat. Sci. For.* **157–162** (1994) 1103.

T. Watanabe, *Mat. Sci. Eng.* **A176** (1994) 39.

T. Watanabe, H. Fujii, H. Oikawa and K.I. Arai, *Acta Metall.* **37** (1989) 941.

T. Watanabe, *Res. Mech.* **11** (1984) 47.

T. Watanabe, *Scripta Met. Mat.* **27** (1992) 1497.

T. Watanabe, *Int. Symp. on Grain boundary engineering*, Ed. U. Erb and G. Palumbo, Canadian Inst. Min. Met. & Petrol., Montreal, (1994) 57.

T. Watanabe, Proc. Conf. Microscale textures of materials, Ed. B.L. Adams and H.J. Bunge, *Tex. & Micros.* **20** (1993) 195.

T. Watanabe, *J. de Phys.* **49** (1988) C5–507.

T. Watanabe, Y. Suzuki, S. Tanii and H. Oikawa, *Phil. Mag. Lett.* **62** (1990) 9.

Y. Yoshimoti, K. Iwayama, T. Nagashima, J. Harase, H. Masui and N. Takahashi, *Mat. Sci. For.* **113–115** (1993) 281.

Y. Yoshitomi, Y. Ushigami, J. Harase, T. Nakayama, H. Masui and N. Takahashi, *Mat. Sci. For.* **113–115** (1993) 715.

J. Zhao, B.L. Adams and P.R. Morris, *Tex. & Micros.* **8/9** (1988) 493.

J. Zhao, J.S. Koontz and B.L. Adams, *Met. Trans.* **19A** (1988) 179.

A.P. Zhilyaev, V.Y. Gertsman, O.V. Mishin, A.I. Pshenichnyuk, I.V. Aleksandrov and R.Z. Valiev, *Acta Met. Mat.* **41** (1993) 2657.

Index

Also from The Institute of Materials

MICROTEXTURE DETERMINATION
and its applications

Valerie Randle

Contents include: basic aspects of microtexture determination; microtexture determination by electron back-scatter diffraction (1) background and general operation and (2) data acquisition; microtexture data processing (1) general principles and (2) practical aspects; application of microtexture to grain boundary studies; outlook and applications.

Book 510   176pp   Hardback   ISBN 0 901716 39 5   £40 (EU)/$80 (non-EU)

For Product Safety Concerns and Information please contact our
EU representative GPSR@taylorandfrancis.com Taylor & Francis
Verlag GmbH, Kaufingerstraße 24, 80331 München, Germany